Cognitive Neuroscience Robotics A

Masashi Kasaki • Hiroshi Ishiguro • Minoru Asada
Mariko Osaka • Takashi Fujikado
Editors

Cognitive Neuroscience Robotics A

Synthetic Approaches to Human Understanding

Springer

Editors
Masashi Kasaki
Graduate School of Letters
Kyoto University
Kyoto, Japan

Minoru Asada
Graduate School of Engineering
Osaka University
Osaka, Japan

Takashi Fujikado
Graduate School of Medicine
Osaka University
Osaka, Japan

Hiroshi Ishiguro
Graduate School of Engineering Science
Osaka University
Osaka, Japan

Mariko Osaka
Division of Cognitive Neuroscience Robotics
Institute for Academic Initiatives
Osaka University
Osaka, Japan

ISBN 978-4-431-56631-1 ISBN 978-4-431-54595-8 (eBook)
DOI 10.1007/978-4-431-54595-8

Cover Photograph by Fabio Dalla Libera

Printed on acid-free paper

This Springer imprint is published by Springer Nature
The registered company is Springer Japan KK

Preface

A variety of technologies are developing and making our society mechanized and computerized at an unprecedented pace, while the overall effects of this development on the human brain are not considered. Some examples of this development include exposure to a massive amount of information via computer networks and the widespread uses of cell phones and automobiles. At least some consequences of the development are not necessarily beneficial. It is highly plausible that our society, as it is today, places an excessive cognitive burden on the brains of children, the elderly, and even adults in the prime of life (see Fig. 1). In order to lead the development of technologies for every member of society in a healthy direction, it is necessary to develop new Information and Robot Technology (IRT) systems that can provide information and services on the basis of the understanding of higher human brain functions (or functions of the "cognitive brain").

In order to deeply understand the human brain functions and develop new IRT systems, Osaka University established the Center of Human-Friendly Robotics Based on Cognitive Neuroscience in 2009, with funding from the Global Center of Excellence (GCOE) Program of the Ministry of Education, Culture, Sports, Science and Technology, Japan. The Center integrates the following world-class research programs and studies at Osaka University, Advanced Telecommunications Research Institute International (ATR), and National Institute of Information and Communications Technology (NICT):

- World-famous human–robot interaction studies: Graduate School of Engineering and Engineering Science, Osaka University, ATR Intelligent Robotics and Communication Laboratories
- Japan's largest-scale program in cognitive psychology: Graduate School of Human Sciences, Osaka University
- World-recognized pioneering studies in brain science and brain machine interface: Graduate School of Medicine, Osaka University, and ATR Computational Neuroscience Laboratories, and NICT

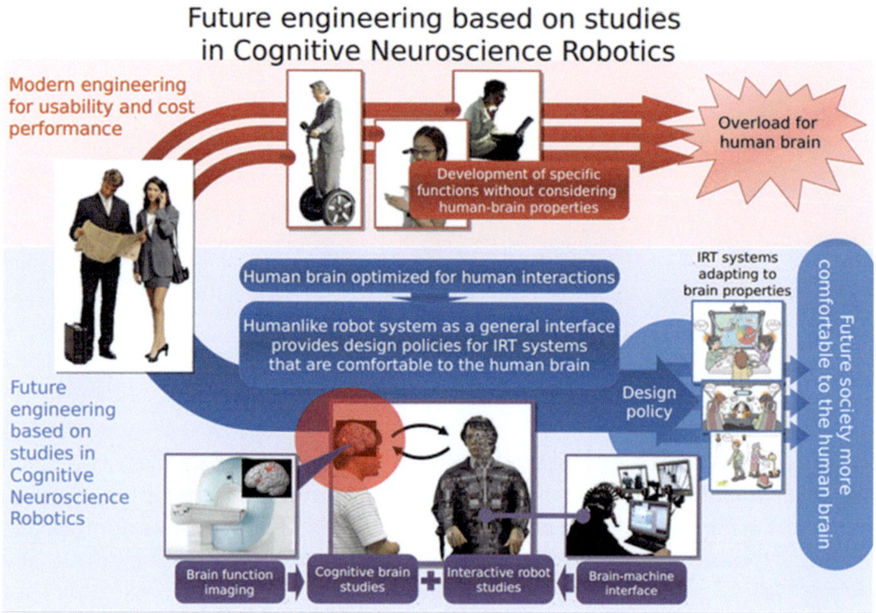

Fig. 1 Traditional engineering vs. future engineering

The Center pursues a new research and education area in which humanities and sciences closely collaborate with each other. Thus, the Center has reorganized education and research at the graduate schools of Osaka University and provided students and researchers a place to engage in the new research and education area. This new area is named cognitive neuroscience robotics.[1]

In more detail, cognitive neuroscience robotics addresses three interrelated research tasks, among others. The first task is to explore how higher brain functions [e.g., consciousness, memory, thinking, emotion, *kansei* (feeling), and so on] are involved in the use of IRT systems, by measuring brain activities with brain-imaging technology. This requires interdisciplinary studies between cognitive and brain sciences. The second related task is to investigate higher brain functions on the basis of brain functional imaging studies on brain function disabilities and studies on brain–machine interfaces (BMI). This requires interdisciplinary studies between brain science and engineering. The third task is to develop prototypes of human-friendly IRT systems and new hypotheses about the cognitive brain by combining studies relevant to the other tasks. In short, cognitive neuroscience robotics, with new technologies at hand, will establish a new understanding of the cognitive brain and develop prototype systems, to solve the problems with modern

[1]The Center finished its proposed research under the funding from the GCOE program in 2014. The Center was then integrated into the Division of Cognitive Neuroscience Robotics, Institute for Academic Initiatives, Osaka University.

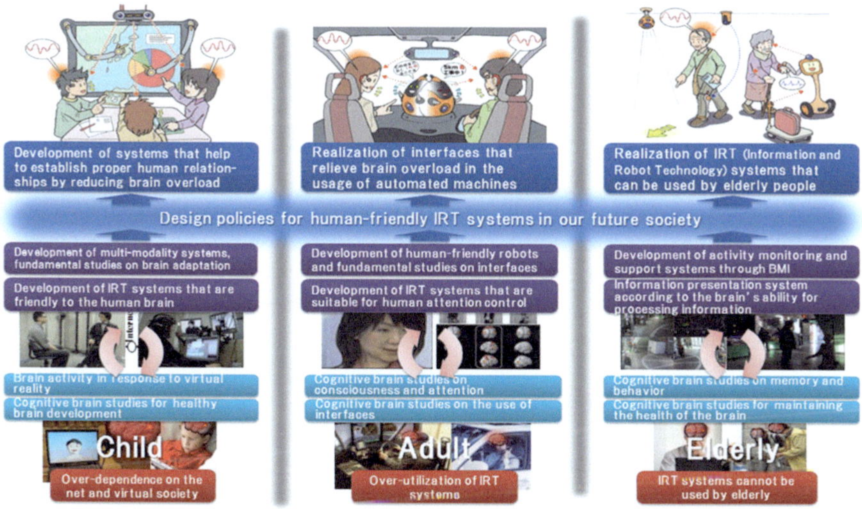

Fig. 2 Solutions by systems based on cognitive neuroscience robotics

mechanized society. Figure 2 shows a typical example of each of the three tasks of cognitive neuroscience robotics.

The Center consists of four interdisciplinary education and research groups. They are organized into a unified five-year education and research program dedicated to the tasks stated above.

- The Group for Establishment of Cognitive Neuroscience Robotics encompasses all research activities in the Center. It establishes the direction of the Center's education and research and aims to systematize the new area of cognitive neuroscience robotics, through scientific and philosophical considerations.
- The Group for Interdisciplinary Studies in Cognitive and Brain Sciences aims to reveal higher brain functions (the cognitive brain) with brain-imaging technology.
- The Group for Interdisciplinary Studies in Brain Science and Engineering develops brain machine interfaces that directly connect the human brain with IRT systems.
- The Group for Development of Cognitive Brain Systems develops prototypes of future IRT systems that do not cause the overload of the human brain, as opposed to the existing IRT systems.

These interdisciplinary research groups include prospective researchers, engineers, and entrepreneurs. The Center offers them a graduate minor program of cognitive neuroscience. This program provides them with basics of cognitive neuroscience robotics and prepares them to address and accommodate the needs of the future society. Students enrolled in the minor program of cognitive neuroscience are required to take two courses: "synthetic approach to human understanding" and "cognitive brain science." Synthetic approach to human understanding and cognitive brain science are two aspects of cognitive neuroscience robotics, seen from the perspectives of robotics and cognitive science, respectively. Each course consists of a series of lectures given by representative researchers in the research groups.

This two-volume book is written as a textbook for prospective researchers in cognitive neuroscience robotics. Volume A, *Synthetic Approaches to Human Understanding, covers the* robotics aspect of cognitive neuroscience robotics and corresponds to the content of the course "synthetic approach to human understanding"; Volume B, *Analytic Approaches to Human Understanding*, covers the cognitive science aspect of cognitive neuroscience robotics and corresponds to the content of the course "cognitive brain science." The chapters of each volume are written by the lecturers of the corresponding course. The two volumes are jointly designed for young researchers and graduate students to learn what cognitive neuroscience robotics is.

We, the editors of this book, strongly hope that you, the reader of this book, will contribute to the development of our society by studying cognitive neuroscience robotics.

Lastly, we would like to convey our appreciation and gratitude to all authors of the individual chapters of this two-volume book. The main editor, Masashi Kasaki, read every chapter and provided detailed feedback to each author. His contribution to the book deserves special mention here.

Japanese Society for the Promotion of Science Postdoctoral Fellow, Graduate School of Letters, Kyoto University
Guest Associate Professor, Division of Cognitive Neuroscience Robotics, Institute for Academic Initiatives, Osaka University
Masashi Kasaki

Leader of the GCOE Center of Human-Friendly Robotics Based on Cognitive Neuroscience
Professor, Graduate School of Engineering Science, Osaka University
Hiroshi Ishiguro

Director of the Division of Cognitive Neuroscience Robotics, Institute for Academic Initiatives
Professor, Graduate School of Engineering, Osaka University
Minoru Asada

Guest Professor, Division of Cognitive Neuroscience Robotics, Institute for Academic Initiatives, Osaka University
Mariko Osaka

Professor, Graduate School of Medicine, Osaka University
Takashi Fujikado

Contents

Main Contributors

Tatsuo Arai Graduate School of Engineering Science, Osaka University, Osaka, Japan

Minoru Asada Graduate School of Engineering, Osaka University, Osaka, Japan

Hiroaki Hirai Graduate School of Engineering Science, Osaka University, Osaka, Japan

Koh Hosoda Graduate School of Engineering Science, Osaka University, Osaka, Japan

Hiroshi Ishiguro Graduate School of Engineering Science, Osaka University, Osaka, Japan

Hiroko Kamide Research Institute of Electrical Communication, Tohoku University, Miyagi, Japan

Takayuki Kanda Intelligent Robotics and Communication Laboratories (IRC), Advanced Telecommunications Research Institute International (ATR), Kyoto, Japan

Takahiro Miyashita Intelligent Robotics and Communication Laboratories (IRC), Advanced Telecommunications Research Institute International (ATR), Kyoto, Japan

Fumio Miyazaki Graduate School of Engineering Science, Osaka University, Osaka, Japan

Yukie Nagai Graduate School of Engineering, Osaka University, Osaka, Japan

Hideyuki Nakanishi Graduate School of Engineering, Osaka University, Osaka, Japan

Yuichiro Yoshikawa Graduate School of Engineering Science, Osaka University, Osaka, Japan

Chapter 1
Compliant Body as a Source of Intelligence

Koh Hosoda

Abstract No one denies that the brain is the main organ for cognition. However, the brain has evolved with the body. We cannot really understand how the brain works unless we understand the function of the body. In this chapter, we review several experimental examples of robots that have similar structures to humans and investigate the function of the body. It turns out that in order for a robot to achieve intelligent behavior, it is extremely important to have a compliant body with a muscular–skeletal structure.

Keywords Trade-off between the stability and controllability • Structural compliance • Biarticular muscle • Underactuated mechanism • Dynamic touch • Proprioceptive sensor • Passive dynamic walking • Coordination between joints (joint coordination)

1.1 Robot Design for Understanding Intelligence

How do we understand that body design is important for emerging intelligent behavior? To understand this, let us first consider two types of aircrafts: a control-configured vehicle (CCV) and a glider. A CCV is designed to enhance motion performance, but it has little aerodynamic stability. It cannot fly without control, but it is highly controllable. For example, it can change flight direction without changing the nose direction. A CCV can fly, because control theory and fly-by-wire technology are well developed. By contrast, a glider is designed for stable aerodynamics. It can fly without any control. However, its flight is totally governed by natural dynamics; thus, its controllability is relatively small. It cannot change the flight direction without changing the nose direction. Nor can it change flight speed so much. There is a trade-off between the stability and controllability of these aircrafts.

The trade-off for these aircrafts is similar to that between a motor-driven walking robot and a passive dynamic walker. When we apply motor control for a walking

K. Hosoda (✉)
Graduate School of Engineering Science, Osaka University, Toyonaka, Osaka, Japan
e-mail: hosoda@sys.es.osaka-u.ac.jp

M. Kasaki et al. (eds.), *Cognitive Neuroscience Robotics A*,
DOI 10.1007/978-4-431-54595-8_1

robot, we first design body motion so that it can walk stably and then derive the desired trajectory for each motor according to the designed body motion. The motor is controlled to track the desired trajectory, and as a result, whole-body motion is realized. Therefore, the whole-body motion of the robot can be easily altered by the designer; its controllability is large. However, the cost in terms of control is very high. Passive dynamic walkers are first introduced by McGeer (1990). Their walking is totally governed by natural dynamics. The dynamics of a walker is designed in such a way that it can walk down (or fall down, so to speak) a shallow slope. Because its walking down is passive and fully based on dynamics, it is difficult to change its walking behavior; however, it can walk without any control cost. Here again, there is a trade-off between stability and controllability.

How about human walking? If a person computes the desired trajectory of each joint and accordingly controls it, necessary computation is enormous. Can all computation be executed in the brain? If so, computation may engage most brain resources, and the person may not be able to execute other tasks while walking. On the other hand, if a person walks in a totally passive way, he/she can only walk down a slope and cannot cope with a flat plane or obstacles. Obviously, actual people walk in a way that incorporates the features of both stability and controllability; while people exploit body dynamics, they control their joints to some extent. In the artificial intelligence field, this is called a "diversity–compliance" or "stability–flexibility" trade-off (Pfeifer and Bongard 2007). We may not be able to understand the human intelligence required for walking if we only focus on either stability or controllability.

We may understand that human walking is neither completely controlled nor completely passive, but can we resolve the trade-off to realize a robot walking as adaptive as human walking? What is the common principle in human and robot bipedal walking? We must control the passivity: we must neither allow the robot to be completely passive nor allow it to control everything. The key idea is regulating *structural compliance*. The human body is compliant in a certain direction and rigid in another direction. A person intentionally controls the directionality of compliance so that he/she prepares for impact. Structural compliance is provided by the human muscular–skeletal and skin structure. By controlling structural compliance, we can shape global passive behavior. This will be the solution for the trade-off. In this sense, regulating structural compliance is very important for understanding human motion intelligence.

The actual human muscular–skeletal structure is very complicated and redundant. The number of control inputs—the number of muscles—is far larger than that of joints. This means that if we only focus on tracking desired trajectories, the number of solutions in terms of muscle excitation patterns is uncountable. This demonstrates the complexity of adaptive intelligence. To understand human motor intelligence, we must deal with the dynamic properties of the human body—the driving mechanism for generating intelligent behavior. In other words, we must focus on the muscular–skeletal system and soft skin. The most direct approach to this complicated system is to build a human-compatible body, let it work in a real environment, and observe how it functions. In this chapter, we will identify several challenges when we try to understand human adaptive behavior by building human-equivalent soft bodies.

Structural compliance is also important for physical sensation of humans and robots. If a human or robot uses vision or hearing sense, the human or robot can observe the environment without physically interacting with it; they can remotely observe the environment. However, if the human or robot uses touch to probe the physical environment, the interaction will involve both action and reaction. If a body is completely rigid, it cannot sense any physical property that can be obtained through touch. (Imagine that a force sensor has a certain elastic element where we can put strain gauges for measuring exerted force. If it is completely rigid, it cannot sense force.) By changing the structure of body compliance, we can control information flow from outside. Thus, the following question arises: how do humans regulate this structured compliance?

In what follows, we first introduce a very important engineering tool for understanding human's embodied intelligence—pneumatic artificial muscles (1.2). Then, we introduce an anthropomorphic robot arm driven by a humanlike muscular–skeletal system (1.3) and demonstrate that body compliance enables dynamic movement (1.4), dynamic touch (1.5), and dexterous manipulation (1.6). We also introduce passive-dynamics-based walking robots (1.7) and demonstrate that body compliance enables dynamic walking (1.8), floor detection (1.9), and dynamic jumping (1.10).

1.2 Pneumatic Artificial Muscles

A pneumatic artificial muscle is an engineering tool for realizing structural compliance equivalent to that of human's. Many prototypes and commercial products of artificial muscle exist. In Fig. 1.1, we show some pneumatic artificial muscles, called "McKibben pneumatic artificial muscles"; each muscle consists of a gum chamber and a covering plastic sleeve. When we supply air through the tube,

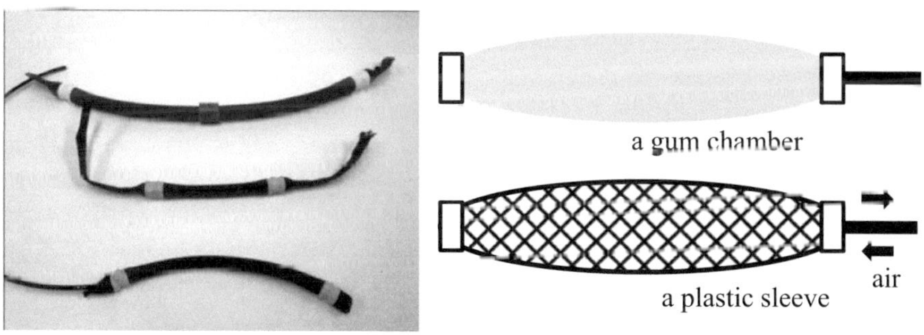

Fig. 1.1 McKibben pneumatic artificial muscles

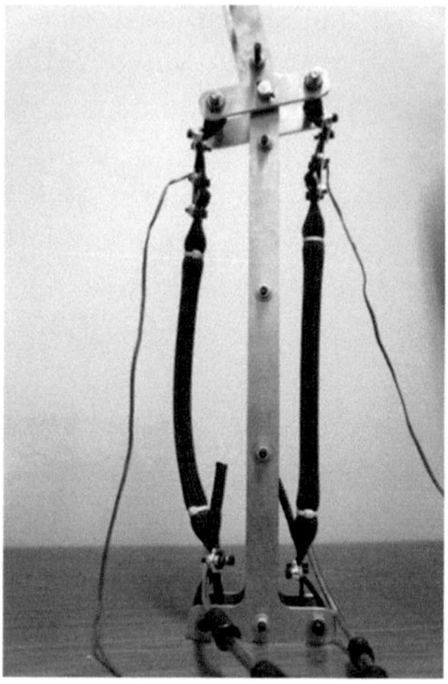

Fig. 1.2 A joint is driven by an antagonistic pair of artificial muscles

the chamber expands like a balloon. The sleeve transfers the expansion force to longitudinal force. Therefore, the pneumatic muscle generates force when air is supplied. Because the working fluid is compressible air and the chamber is made from gum, the artificial muscle is essentially compliant. We will come back to the point that compliant muscles are important for biological systems including humans to achieve adaptive behavior.

A certain time is required for the working fluid to flow into the chamber, and thus, there is a significant time delay between the opening of the valve and force generation by the artificial muscle. In addition, the friction between the gum chamber and the plastic sleeve causes nonlinear hysteresis. For these two reasons, it is very difficult to control an artificial muscle. It was a trend in the 1980s to control highly nonlinear artificial muscles by a nonlinear control method, such as neural network. However, all such attempts have failed due to the complexity of the characteristics of the pneumatic muscle. Pneumatic artificial muscles are compliant but not really controllable; for them, again, there is a trade-off between compliance and controllability.

If a joint is driven by an antagonistic pair of pneumatic artificial muscles (Fig. 1.2), we can change joint compliance by increasing or decreasing the amount of air for them. This mechanism is very similar to the mechanism of humans; in this mechanism, compliance plays a large role in realizing an automatic reaction when a torque is exerted on a joint. If a joint is controlled by a motor, we have to sense the position and velocity of the joint and the torque applied to the motor; we also have to feed the information on these back to the motor to realize the reaction. Each of these requires computation and time. We can expect a humanlike automatic reaction if we apply pneumatic artificial muscles. Note that, again, there is a trade-off between compliance and diversity.

Because McKibben pneumatic artificial muscles are compliant not only along the longitudinal direction but also along the radial direction, we can build a humanlike

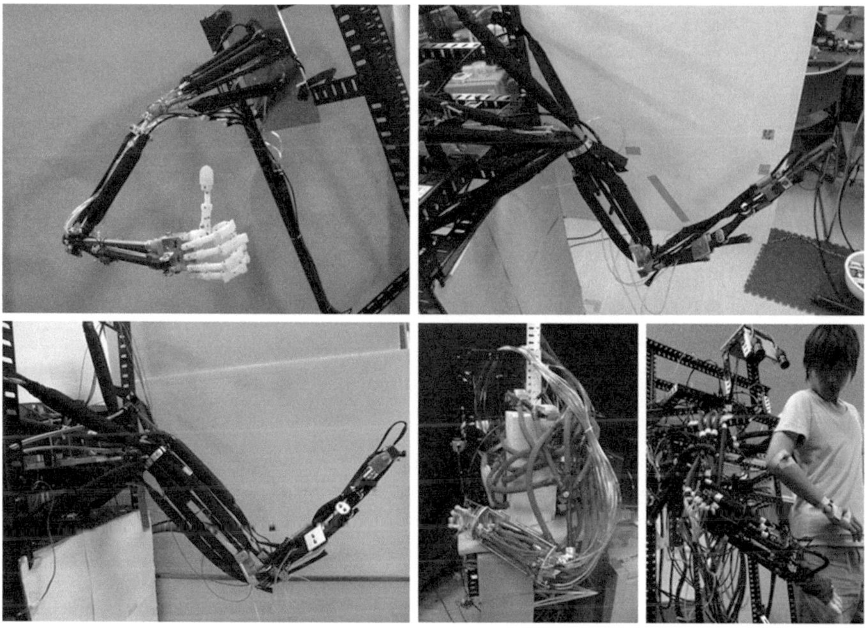

Fig. 1.3 Muscular–skeletal robots developed at Osaka University

muscular–skeletal structure. The robotics group at Osaka University has developed many anthropomorphic humanoid robots to understand the role of body structure in intelligent behavior (Fig. 1.3). First, let us look at an anthropomorphic robot arm.

1.3 Anthropomorphic Robot Arm

An anthropomorphic robot arm that has a muscular–skeletal structure similar to that of a human helps us to understand the adaptive motion of humans (Fig. 1.4) (Hosoda et al. 2012). By using the robot to complete human tasks, we may understand the functions of the structure and its contribution to intelligent behavior. First, we need to try to imitate the human arm as much as possible without considering the tasks that the robot will accomplish. Actually, robots have traditionally been designed to achieve given tasks very well. Even today, most robots are designed to have certain purposes before they are built. For this reason, traditional robots are very good at performing predefined tasks but typically not capable of performing any other tasks; their abilities are very limited even though they are supposed to be versatile. On the other hand, we wish to design an anthropomorphic robot arm to help us estimate how much intelligence is embedded in the human body. Therefore, we must design a robot without considering specific tasks for it. Interestingly, the trade-off between limitation and versatility in task is, again, a stability and controllability trade-off.

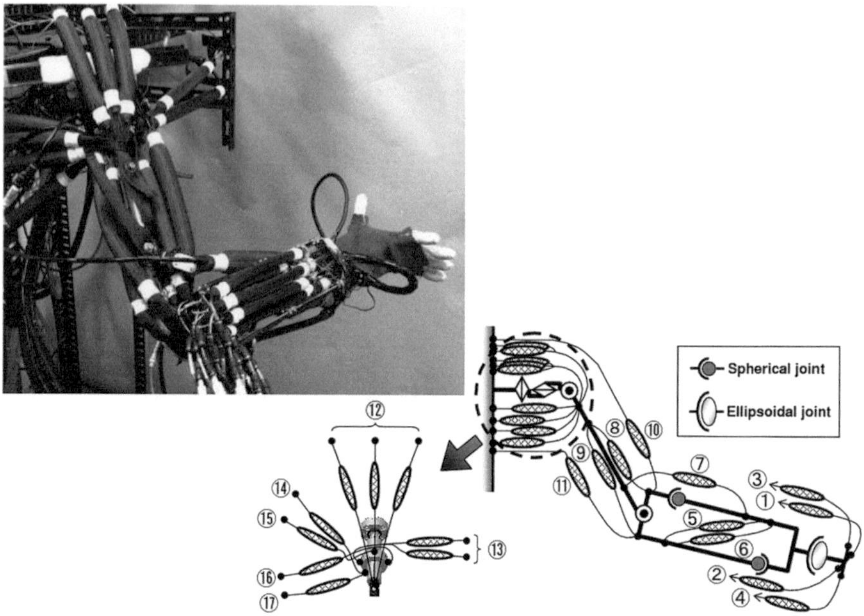

Fig. 1.4 Structure of an anthropomorphic robot arm. The muscles of the robot emulate these human muscles: (1) flexor carpi radialis, (2) extensor carpi radialis, (3) extensor carpi radialis (radial flexor), (4) extensor carpi ulnaris (ulnaris flexor), (5) forearm pronator, (6) forearm supinator, (7) brachioradialis, (8) brachialis, (9) anconeus, (10) biceps, (11) triceps, (12) deltoid, (13) greater pectoral, (14) musculus subscapularis, (15) latissimus dorsi, (16) infraspinous, and (17) musculus teres minor

On the basis of knowledge of human anatomy, we can pick some major muscles in the human arm and design an anthropomorphic robot arm. We must account for the limitation of the engineered parts—typically, the limitations of McKibben pneumatic artificial muscles with regard to contraction ratio and generated force. We can replicate the structure of the human forearm almost as it is in the human body. For example, the forearm consists of two bones—radius and ulna—and humanlike pronation and supination can be realized. Traditional robots driven by electric motors cannot realize this. The robot's wrist is also designed to emulate the function of the human wrist. The human wrist consists of small bones and functions as a semisphere. It can bend and swing the hand, and pronation and supination can be realized by radius and ulna. This is a very typical human motion and cannot be realized by motor-driven robots. The human shoulder is too complicated and is hardly realized by existing engineered parts as it is. Therefore, we focus on the functional joints of the shoulder and realize it by a ball joint and several muscles.

The human arm has seven degrees of freedom (DOF): three DOF for the shoulder, one DOF for the elbow, and three DOF for the wrist. The anthropomorphic robot arm also has 7 DOF and is driven by 17 muscles. As we mentioned above, the least number of muscles to drive one joint is two, which means that 14 muscles

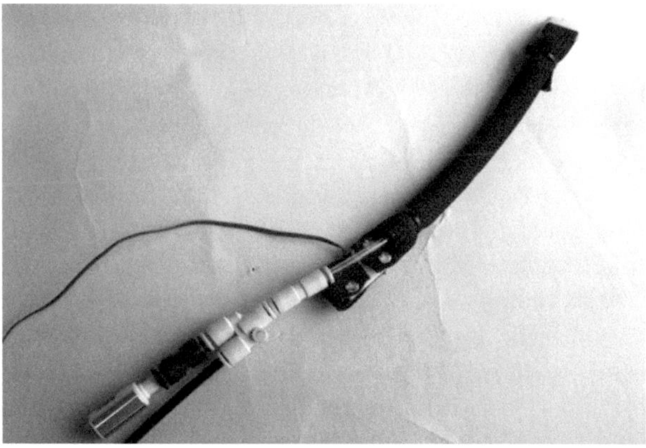

Fig. 1.5 Sensors for a McKibben pneumatic artificial muscle, a pressure sensor, and a tension sensor

are sufficient to control an arm with 7 DOF. This implies that our robot arm includes redundancy (more inputs than required for the number of controlled objectives). In addition, muscle 10 (biceps) drives not one but two joints, making it a "biarticular muscle." For these reasons, the whole muscular–skeletal system is very complicated.

We cannot use normal angle sensors, such as encoders and potentiometers, for the anthropomorphic robot, because each joint does not have a fixed axis of revolution. Humans have proprioceptive sensors, such as Golgi tendon organs and muscle spindles in the muscles, and they can sense the state of muscles. We can put a tension sensor and a pressure sensor on a McKibben pneumatic artificial muscle, and they sense the state of the muscle; this is similar to how a natural muscle works (Fig. 1.5). The length of the muscle can be estimated from tension and pressure (Chou and Hannaford 1996).

The anthropomorphic robot arm has a hand that can be used to grasp objects. The hand can be designed to precisely imitate the human hand for a more detailed investigation into human function; however, for this robot, we reduce the number of muscles by utilizing an underactuated mechanism so that we can focus on the mechanisms of the arm.

1.4 Dynamic Motion of the Anthropomorphic Robot Arm: Throwing

When a body is driven by antagonistic pairs of pneumatic artificial muscles, the joint can either be very stiff, compliant, or totally soft, depending on the amount of supplied air. By utilizing this feature of the joint, the robot can regulate its dynamic property. For example, if the air is totally exhausted, every muscle is very soft, and the robot can swing the arm naturally. If some muscles are partially supplied with air, the robot can utilize the compliant muscles to rapidly move from one point to another. This can allow dynamic actions, such as throwing. It is quite difficult for a traditional robot with gears driven by electric motors to achieve quick motions, such as throwing. Speed must be reduced to increase the torque, and thus, the robot does not move fast. To realize fast motions, the robot needs to be driven by bulky direct-drive motors, or friction needs to be reduced or compensated for, using torque sensors.

Braun et al. (2013) demonstrate that the elasticity of an actuator is beneficial for throwing, and they use series elasticity with electric motors. So far, no study has identified how to use compliance with humanlike muscle-and-bone structures. The muscular–skeletal system is, as mentioned above, very complicated, and it is very difficult to control it precisely. The point is to create a robot with dynamics similar to those of the human muscular–skeletal structure. We can directly teach the robot if it has similar dynamics; the instructor should be able to say, "now, move your elbow forward" and "now, stretch your elbow" as if the instructor teaches a human, and an intermediate motion should be naturally generated by the dynamics. Figure 1.6 shows an example of robot arm capable of throwing a ball; the designer

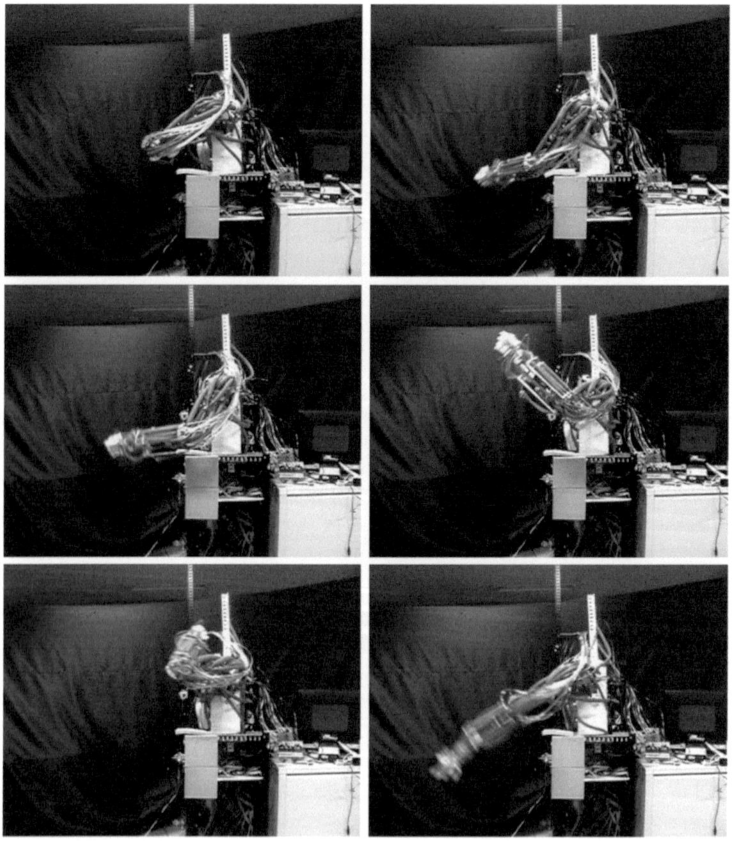

Fig. 1.6 A sequence of throwing a ball by the anthropomorphic robot arm (snapshots in every 0.4 s)

has only programmed several representative postures by a certain valve pattern, and an intermediate motion is interpolated by the robot's dynamics. The pictures are snapshots at 0.4-s intervals. We can see the very humanlike throwing motion of the robot.

The open or close pattern of the valve used in this experiment is found by the designer in a top–down manner; however, it is interesting to note that the motion of the robot looks very similar to that of a real human, perhaps because the muscular–skeletal structure of the robot is similar to that of a real human. We should expect that the robot will generate a natural humanlike motion without a program for the precise motion of each joint.

The robot utilizes its compliance to store energy and releases it in a short period of time to throw a ball. This is a ballistic control process in which the robot is controlled only at the beginning. After the robot is driven for a certain period of time, it is no longer driven but moves freely; actually, this is not completely *free* and its moves are subject to its dynamics. Since the throwing motion is fast, it is not realized by precise motion control—even if that is possible, it will be energy consuming.

1.5 Sensing Exploiting Compliance: Dynamic Touch

We humans sometimes use an exploring strategy, called "dynamic touch," to acquire the dynamic properties of a grasped object (Hosoda et al. 2012). Dynamic touch allows us to estimate certain properties of an object, such as weight, moment, and compliance, by shaking it (Fig. 1.7, left). Consider the use of dynamic touch by a traditional robot (Fig. 1.7, right). The traditional robot arm is driven by electric motors and gears. Its joints are stiff and cannot be bent by inertial force of the object. Therefore, even if the arm shakes an object, its behavior does not change, whether the object is heavy or light. The robot cannot acquire information about the object by observing the joint motion. To get this information, we have to put

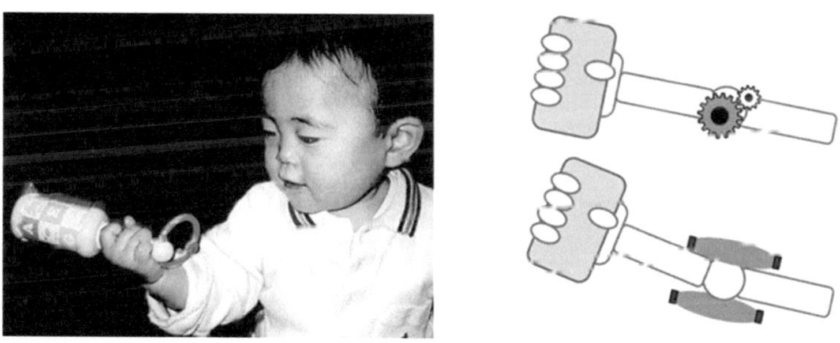

Fig. 1.7 Dynamic touch. Dynamic touch is difficult for motor-driven robots but easy for compliant robots that are driven by antagonistic pairs of muscles

either a force/torque sensor in the wrist or a torque sensor at each joint of the robot. The sensor will lose rigidity when it gains sensitivity; there is a trade-off between rigidity and sensitivity. For a traditional robot, rigidity is equal to precision, and it is of the greatest importance. Therefore, even if the sensitivity of the sensors can be increased, it leads to a great increase in cost.

Now, let us look at the dynamic touch of humans. A human has a muscular–skeletal system and soft skin. When a human performs dynamic touch on an object, the object is shaken and the reaction force from the object is applied to the muscles. Then, the proprioceptive sensors can get information about dynamics of the object. In addition, there are many receptors in the soft skin. While the object is being shaken, these receptors provide an enormous amount of information about the object. The soft skin not only acts as a sensor system but also provides cushioning. This allows a human to grasp the object firmly. By comparing the robot's dynamic touch with the human's, we can see that sensory data are quite expensive for traditional "rigid" robots, whereas humans can use their softness to learn more about an object.

We can investigate the performance of dynamic touch by a real muscular–skeletal robot. The anthropomorphic robot arm introduced in Sect. 1.3 has tension and pressure sensors, and they work as proprioceptive sensors for artificial muscles. The robot hand has soft skin; it has several strain gauges inside the skin. We measure the extent to which the robot can categorize an object on the basis of information from these sensors. It is relatively obvious that the robot can detect the weight and moment of an object. Therefore, we should also look at more detailed information about dynamics.

We have performed an experiment with the goal of having the robot not only estimate the weight of a plastic bottle but also detect the content of the bottle: liquid or solid. If the content is solid, the robot should detect whether the bottle contains flour, sand, or stones. If the content is liquid, the robot should detect whether the bottle contains water or oil. The robot can change the way in which it shakes the bottle—horizontally or vertically—and identifies the content on the basis of its proprioceptive sensory signals.

Before conducting the experiment, we asked human subjects to perform the same test. We found that the subjects made their guesses mainly on the basis of sound. The task was very difficult for them when they used earplugs. In the experiment, the robot does not use any sound information.

Figure 1.8 shows the experimental results in which the robot estimates the weight of an object. Because the arm robot has 7 DOF, it can shake the bottle in any direction. Here, we show two types of shaking: horizontal and vertical shaking. The robot can change the speed at which it completes one shake of the bottle. The figure shows the results when the robot has shaken the bottle (a) vertically at 1 Hz, (b) vertically at 1.25 Hz, and (c) horizontally at 1 Hz. The number of the dimensions of data that we obtain from the proprioceptive sensors and skin sensors is large. Therefore, when the robot categorizes the object on the basis of sensory data, it can utilize multiple dimensions of data. To help readers, we depict the data points in two

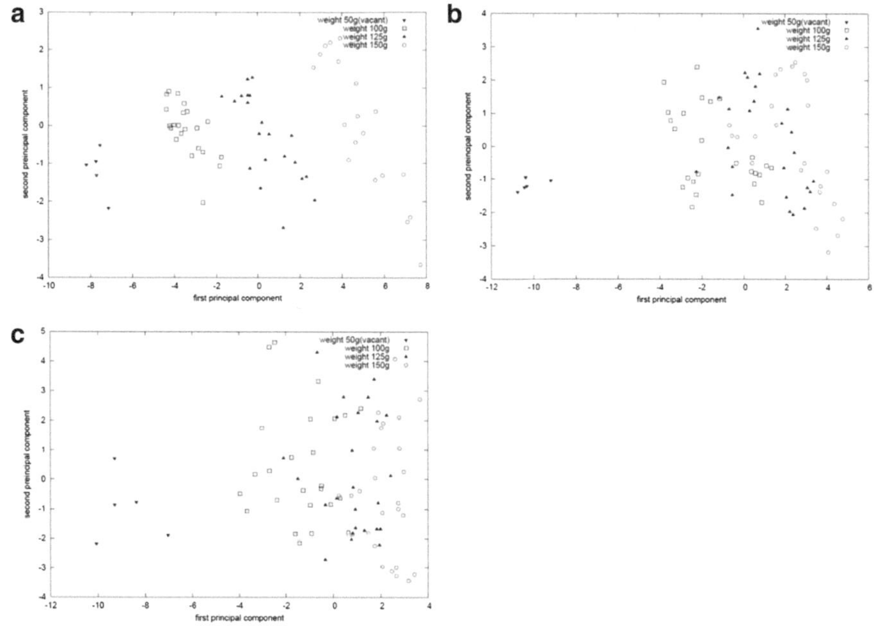

Fig. 1.8 Results of shaking experiments of identifying the weight of a bottle (This figure and caption are modified from Hosoda et al. (2012), with kind permission from Taylor & Francis). (**a**) Shaking vertically at 1 Hz. (**b**) Shaking vertically at 1.25 Hz. (**c**) Shaking horizontally at 1 Hz

dimensions by applying principal component analysis. In the figure, the horizontal and vertical axes are the first and second principal components, respectively.

We can see from Fig. 1.8 that the robot can estimate the weight of a bottle when it shakes the bottle vertically at 1 Hz. The first component (horizontal axis) indicates clear categories of the weight, but the second component (vertical axis) does not, maybe because of the direction of the gravity. When the robot shakes the bottle at 1.25 Hz or when it shakes the bottle horizontally, it cannot clearly identify the weight. It is quite difficult to explain why these interesting results obtain; we can only conclude that shaking vertically provides more information about weights than shaking horizontally. This may be related to the eigenfrequency of the system; however, it is quite difficult to model all the physical phenomena involved in this experiment.

Inverted triangle, square, triangle, and circle points are data points for empty bottles, 100-g bottles, 125-g bottles, and 150-g bottles, respectively, regardless of whether they contain water, oil, flour, sand, or stones. Different contents may produce different touch sensations; however, the shaking of the bottle does not provide enough information for the information processing system to detect the differences among the contents.

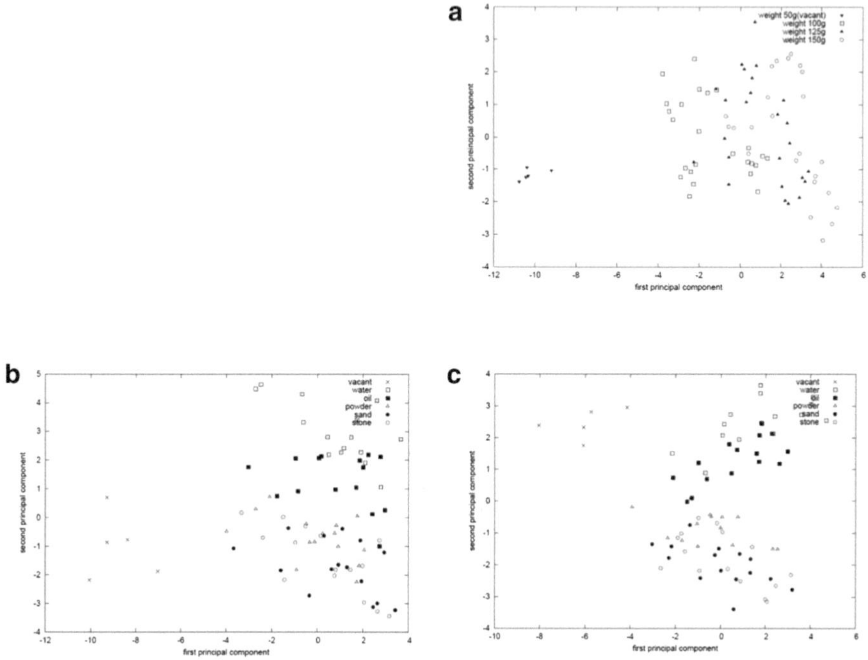

Fig. 1.9 Results of shaking experiments of distinguishing the contents of bottle (This figure and caption are modified from Hosoda et al. (2012), with kind permission from Taylor & Francis). (**a**) Shaking vertically at 1.25 Hz. (**b**) Shaking horizontally at 1 Hz. (**c**) Shaking horizontally at 1.2 5 Hz

In the second experiment, the robot attempts to identify the content of the bottles. Figure 1.9 shows the experimental results when the robot shakes the bottle (a) vertically at 1.25 Hz, (b) horizontally at 1 Hz, and (c) horizontally at 1.25 Hz. These results are basically the same as those in Fig. 1.8 and focus on identifying water (white squares), oil (black squares), flour (white triangles), sand (black circles), and stones (white circles). When the robot shakes the bottle vertically, the data points are scattered almost randomly; but when the robot shakes the bottle horizontally at 1.25 Hz, liquid (water and oil) can be distinguished from solid (powder, sand, and stones) on the basis of the second component (vertical axis). This result is different from the result of the first experiment. It is unclear what mechanism lies behind this difference. These experimental results are highly dependent on the dynamics of the anthropomorphic robot arm. We will get a different result from another compliant robot arm. The conclusion here is not, of course, that this type of experimental protocol is always valid but that we can utilize proprioceptive sensors to categorize objects on the basis of dynamic touch because a change in object dynamics changes robot behavior. The robot can learn to distinguish among objects through experience.

1.6 Manipulation of the Door

A different task should be used to demonstrate the versatility of the anthropomorphic robot arm. An object manipulating task might be used, but it would be too simple unless certain physical constraints are imposed. In a door-opening task, the door provides certain constraints with respect to the ground (Fig. 1.10). The door has directional constraints: it can move along the rotational axis, but cannot move in the perpendicular directions. To perform this task, a traditional rigid robot manipulator would need detailed descriptions of the door, itself, and the relationship between them. With such descriptions at hand, it would apply hybrid position/force control in order to open the door without breaking it.

The movement of the door is constrained by the hinge, as well as the ground: it only moves along an arc. If the robot grasps the knob, the robot needs to deal with a very complex relation to it; the robot should change the position and rotation of its hand according to the movement of the knob. A small change in hand position and rotation may lead to a large change in reaction force, which may break the robot or the hinge, or both. The control cost will be enormous.

When an anthropomorphic robot arm opens a door, it does not generate too much force even if relative position and orientation of the door with respect to the robot changes. This is because the robot is essentially compliant. It is not only compliant but also has structured compliance that is expected to be similar to that of humans. Therefore, even if there is a position or orientation error in the relationship between the robot and the door, the pattern of the reaction force generated by the error is similar to the reaction force by a human when there is a position or orientation error in the relationship between a human and the door. Imagine that the knob is placed high on the door, and we have to stretch to grab it. The anthropomorphic robot arm *feels* the same reaction force in each muscle. This is quite different from the reaction

Fig. 1.10 The anthropomorphic robot arm opens a door

force pattern generated by a robot that has series elasticity in each electric motor, since the way of reaction against the external force is quite different. The force pattern of the robot is represented by a Jacobian matrix (Yoshikawa 1990), and the matrix of the anthropomorphic robot is quite different from that of the traditional robot. The usefulness of a tool consists not only in its kinematics but also in the reaction force generated by the environment. In other words, if a robot has the same muscular–skeletal structure as a human does, they both find the same tools useful.

A human learns to use less and less muscular force to complete a task from practice and thus reduces the amount of energy required for completing the task. A robot with a similar muscular–skeletal structure can do the same. To return to the door-opening task described above, if the knob turns smoothly, humans try to turn it with pronation/supination only. If their relative position is not suitable, humans can try to move their body to use fewer muscles instead of repeating the same behavior with many muscles. By building a robot that has the same muscular–skeletal structure as a human does, we can learn more about natural human behavior.

1.7 Muscular–Skeletal Robot Legs

We have discussed on an anthropomorphic robot arm and seen how the structured compliance can be utilized for intelligent behavior. It is natural to extend this discussion to include legged robots. Muscular–skeletal robot legs can be expected to achieve dynamic and adaptive locomotion, such as walking and jumping, by exploiting their own dynamics; this is very difficult for traditional robots because they are controlled to track predefined trajectories. However, the design principle for a legged robot is quite different from that for a robot arm; legged robot is not fixed to the ground and needs to deal with its stability issues before (or simultaneously) realizing meaningful behavior. If we start by imitating the muscular–skeletal structure of the human leg as precisely as possible—as we did for the robot arm—the robot will have great difficulty in keeping its balance. Balance is a fundamental and essential problem for legged robots. The problem cannot be solved by randomly exciting each muscle, as we did for the robot arm.

For dynamic balance, therefore, we start with passive dynamic walking (McGeer 1990). A passive dynamic walker is analogous to a rimless spoked wheel rolling down a slope (Coleman et al. 1997). While it rolls down the slope, its velocity increases and its potential energy decreases. The increment of kinetic energy is dissipated at impact. It can *walk* stably by balancing these energies. A passive dynamic walker can walk down a slope without any actuation or control. But it cannot walk on a flat plane, because it cannot supply energy to recover the amount of energy that is dissipated at impact. To keep the robot walking, we must add actuators to supply energy. We need actuators: they can drive joints when they supply energy, and they can free joints when the robot swings the legs. Electric motors with gears provide the most popular solution for energy supply to the robot; however, gears normally have a significant amount of friction and are not suitable

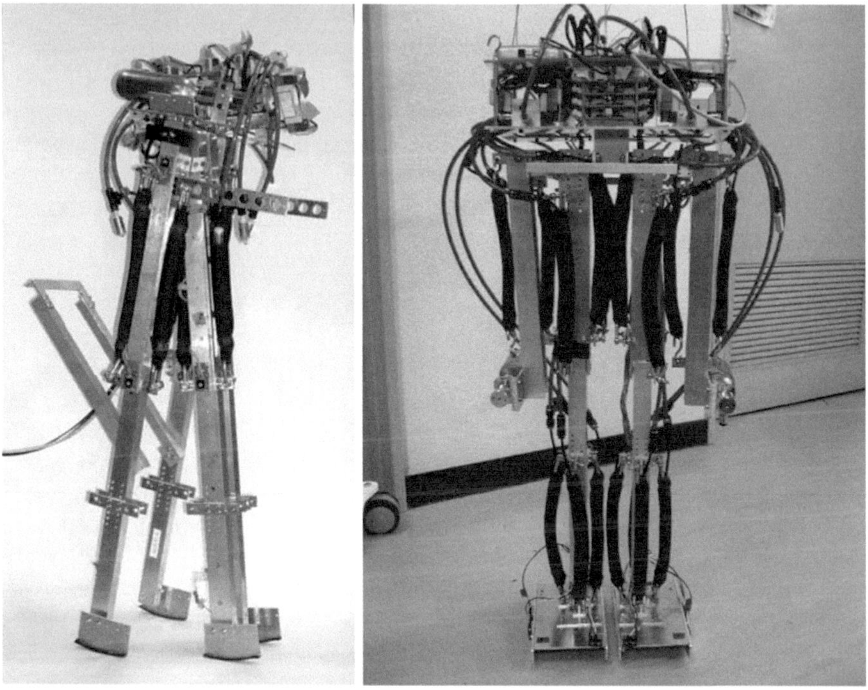

Fig. 1.11 Muscular–skeletal walking robots based on passive dynamic walkers. The *left* walker is a 2D walker without an upper torso. The *right* walker is a 3D walker with a small torso and two arms (This figure and caption are modified from Takuma and Hosoda (2007a), with kind permission from Fuji Technology Press)

for freely swinging the legs. We can use direct-drive motors with less friction, but they are heavy and bulky.

Antagonistic drives powered by pneumatic artificial muscles are suitable for supplying energy to passive dynamic walkers—sometimes they can drive the joints, and sometimes they can let the joints swing. Because they are very light, they do not increase the weight of the leg excessively. Several types of walking robots have been developed on the basis of the idea of adding pneumatic artificial muscles to passive dynamic walkers. Figure 1.11 (left side) shows a 2D walker without an upper torso, and it is built by adding pneumatic artificial muscles to an original 2D passive dynamic walker (McGeer 1990). Figure 1.11 (right side) shows a 3D walker with two arms. Each arm is physically connected to a leg on the opposite side (the left arm to the right leg and the right arm to the left leg). This robot has actuated ankles and so can maintain frontal balance without explicit control, but basically, it is also a passive dynamic walker driven by pneumatic artificial muscles.

These two robots inherit some features of passive dynamic walkers: for example, they have round foot soles, they have knee stoppers to avoid hyperextension, the majority of the weight is placed at the top of the body, and so on. These robots cannot be controlled using traditional controllers that track a predefined trajectory; but they are controlled ballistically. For ballistic control, robots need to know when the swing leg touches the ground. Hence, the two robots are equipped with touch sensors on the foot soles; however, they have no other sensors, such as gyroscopes or accelerometers, which are common sensors for walking biped robots. The success of the robots demonstrates that controls for a walking robot can be simple when the robot body is appropriately designed (Pfeifer and Bongard 2007).

1.8 Ballistic Walking Control

Ballistic walking control is periodical: after one leg touches down, the other leg (the swing leg) is driven for a certain period of time and then freed to swing forward until it touches the ground. Wisse and van Frankenhuyzen (2003) attached pneumatic artificial muscles to a passive dynamic walker and achieved stable walking by applying a ballistic controller, in which they supplied air to the muscle that drove a swing leg after a certain period of time—say, 0.3 s—and let the leg swing forward according to the inertial force. The trajectory of the swing leg is, of course, not designed in advance. This type of control is called "phase-based control" or "limit-cycle control," and it is demonstrated to be relatively stable; for example, the Wisse and van Frankenhuyzen robot can walk outside.

Note that ballistic control does not explicitly specify a walking cycle. The trigger is the signal from the touch sensor in the sole. After the touch, the robot supplies the muscle with air that drives the swing leg. The leg swings forward, and the gravity center also moves forward. The whole body leans forward, and the swing leg touches down. This cycle of movements is not explicitly *programmed*. The body dynamics *knows* how to move. Hence, body dynamics replaces some amount of computation. This is called "morphological computation." Morphological computation is presumed to be essential for the intelligence of autonomous agents (Pfeifer and Bongard 2007).

In a way that is similar to the glider's behavior, passive dynamic walking and ballistic walking are governed by body dynamics, and the cycle of movements is not explicitly programmed. Of course, the speed changes when the angle of the slope changes, but how can we change the walking velocity intentionally without changing the environment? The answer is by altering the dynamics of the robot by changing the compliance of the body.

Let us return to the idea that a robot is antagonistically driven by artificial muscles. Because each joint is antagonistically driven, not only equilibrium angle but also joint compliance can be controlled. The eigenfrequency of the robot is determined on the basis of balance between inertia (mass) and compliance. By changing compliance, the robot can change its walking speed (Takuma and Hosoda 2007b). If the walking speed is not the problem that matters, the robot only has to contract the driving muscle for walking (Wisse and van Frankenhuyzen 2003);

however, additional contraction of the antagonistic muscle obviously changes the swing speed, and the robot can change its walking speed.

Think again about a motor-driven robot. To change its walking (body) speed, it must recalculate joint trajectories and apply joint control on the basis of the recalculated joint trajectories. The walking speed is given in a top–down manner; therefore, we are prone to fix the walking (body) speed at a certain constant level, if we do not take the dynamics of the robot into account. When we observe human walking, we can see that the body speed is not really constant—the speed fluctuates more or less as the body swings back and forth, perhaps because this is energy efficient. Maintaining a constant body speed does not necessarily lead to mean energy efficiency.

On the other hand, if the robot changes its walking speed by changing compliance, it cannot achieve an explicitly specified velocity. The velocity is determined by an interaction between the body and the environment. However, because the change in walking speed is based on robot dynamics, we can expect it to be energy efficient. By introducing this ballistic walking methodology, we can establish a new paradigm in robot walking research and hopefully explain natural human walking control. In this context, the compliance of joints realized by the antagonistic drive is very important.

1.9 Sensing the Terrain by Walking

As we discussed above, an anthropomorphic robot arm can achieve an interesting type of information processing and measurement, such as dynamic touch based on compliance provided by muscular–skeletal mechanisms and soft skin. Robot legs can be also expected to have sensory ability based on compliance.

If a robot is rigid and controlled to follow the desired trajectory without considering the force from the terrain, the movement of each joint does not change even if the terrain changes. The robot falls down easily, because it does not change its behavior so as to reflect terrain changes. To avoid this consequence, a rigid robot usually has force and torque sensors on its feet. These sensors acquire changes in reaction force. In addition, the gyroscopes and accelerometers sense attitude changes. The robot is controlled to be compliant on the basis of information from these *external* sensors.

By contrast, if a robot is essentially compliant and controlled ballistically, changes in terrain dynamics change whole body movement. The movement of each joint is affected by changes in terrain. Therefore, if the robot can continue to walk, it can sense changes in terrain even with proprioceptive sensors. It does not need any external sensors to observe the outside world.

In Fig. 1.12, we show the walking cycle when the 2D legged robot (Fig. 1.11, left) walks on two different terrains: linoleum and carped. For both terrains, we have applied the same controller and same control parameters, such as duration for air supply for the swinging leg. We can see that the robot changes its walking

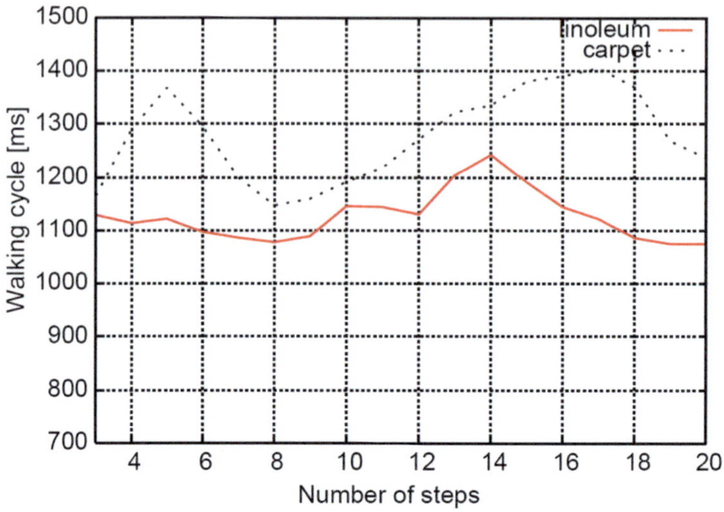

Fig. 1.12 Comparison of the walking cycle of a 2D legged robot driven by pneumatic artificial muscles when it walks on a linoleum and carpet (Takuma and Hosoda 2007a). Even if the same ballistic control is applied, the robot changes its behavior on the basis of terrain dynamics (This figure and caption are modified from Takuma and Hosoda (2007a), with kind permission from Fuji Technology Press)

behavior on the basis of changes in terrain, even if the same controller is applied. The differences between the walking cycles stems from the differences in the combined dynamics of the robot and terrain.

One interesting aspect of the robot's walking behavior is that the robot can learn about a terrain by walking on it. A traditional robot first observes the terrain, plans the trajectory, and finally acts, i.e., walks on the terrain. This sequence of activities is called the "Sense–Plan–Act" methodology, a typical approach of the traditional artificial intelligence (Pfeifer and Bongard 2007). It is known that if we take such methodology, required computation tends to be large, and the cycle time for control becomes long. As a result, the system is brittle against change of the environment. By contrast, a compliant robot driven by muscular–skeletal mechanism first walks, senses the terrain by observing its own behavior, and changes the plan, which makes the robot adaptive and robust without formal modeling of the robot and environment.

1.10 Stable Hopping by a Muscular–Skeletal Mechanism

We have only discussed about monoarticular muscles and joint compliance that are important for realizing ballistic walking and terrain sensing, after Sect. 1.7 (Hosoda et al. 2010). All the joints of the robots in Fig. 1.11 are driven by antagonistic pairs of monoarticular artificial muscles. Now, we return to the structure of the human body and gradually increase the complexity of the robot so that we can imitate the

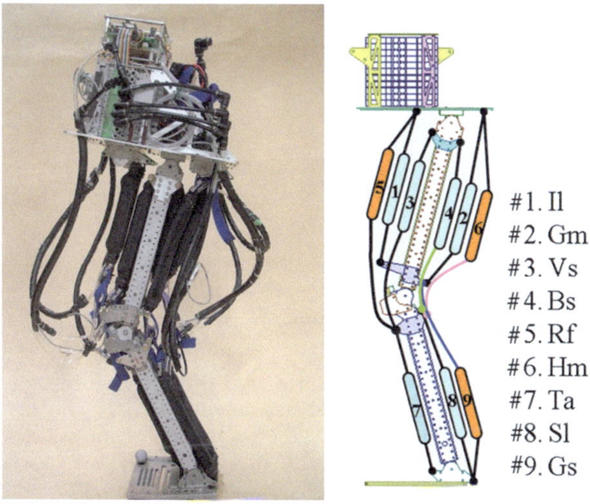

Fig. 1.13 A jumping robot with anthropomorphic muscular–skeletal structure (Hosoda et al. 2010). Muscles #1 (iliacus) and #2 (gluteus maximus) are monoarticular muscles that drive the hip joint. Muscles #3 (vastus lateralis) and #4 (popliteus) drive the knee, and muscles #7 (tibialis anterior) and #8 (soleus) drive the ankle joint. Muscles #5 (rectus femoris) and #6 (hamstring) are biarticular muscles that drive hip and knee joints. A muscle #9 (gastrocnemius) is a biarticular muscle that drives knee and ankle joints (This figure and caption are modified from Hosoda et al. (2010), with kind permission from Springer Science and Business Media)

muscular–skeletal structure of the human. A human has biarticular muscles that drive more than one joint, as we saw in the arm. These biarticular muscles are regarded to contribute to the coordination of joints (Jacobs et al. 1996) and the force transfer from a proximal part to a distal part. This is how animals can reduce the weight of the distal part, and it is very important for fast motion and energy efficiency, as we will discuss later.

Figure 1.13 shows an anthropomorphic robot that imitates the major muscles of the human leg in the sagittal plane. The names of human muscles are indicated on the right side, and the actual robot is shown on the left side. Muscles #5, #6, and #9 are biarticular muscles.

Muscles #5 and #6 are antagonistic and connect the shank with the torso over two joints: waist and knee joints. These muscles contribute to the coordination between these two joints and to the power transfer from the waist joint to the shank. These phenomena are very important for biological systems. A biological system may have large muscles in the proximal part of the body, and power is transferred by the biarticular muscles to the distal part of the body. The transfer of power by the biarticular muscles can reduce the weight of the distal part of the body, which, in turn, can significantly reduce energy consumption. If a large mass is located at the distal part of the body, driving energy is large; this is not beneficial for survival. It is also related to the eigenfrequency of the leg. If the distal part is heavy, the eigenfrequency becomes low, which is a large disadvantage for fast motion.

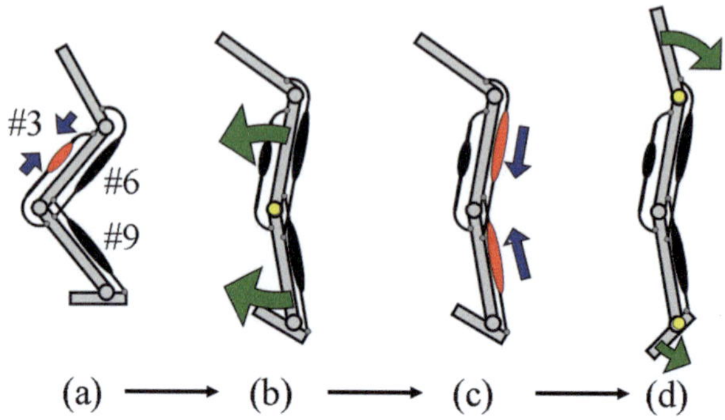

Fig. 1.14 Joint coordination driven by knee extensor #3 (Hosoda et al. 2010): (**a**) when extensor #3 contracts, (**b**) the knee is extended, (**c**) biarticular muscles #6 and #9 transfer force to hip and ankle joints, respectively, and (**d**) the whole leg is extended (This figure and caption are modified from Hosoda et al. (2010), with kind permission from Springer Science and Business Media)

Biarticular muscle #9, gastrocnemius, drives knee and ankle joints. This muscle is also interesting from the mechanics viewpoint. It is not antagonistic to any muscle. We can conjecture that it is an evolutionally obtained body structure advantageous for forward moving, together with the knee locking mechanism. But the exact mechanics have not yet been fully revealed.

Let us look at the joint coordination driven by knee extensor #3 (Fig. 1.14). When it contracts, the knee is extended (Fig. 1.14b). The extensor pulls biarticular muscle #6, and it extends the waist joint. The extensor also pulls biarticular muscle #9, and it extends the ankle joint (Fig. 1.14c, d). As a result, simply by contracting the knee extensor, the whole body is totally extended; this is obviously beneficial for jumping. When the robot touches the ground after a flight phase, the ankle is flexed, and gastrocnemius (#9) is pulled, causing the knee to flex. The flexed knee pulls muscle #5 and it flexes the waist joint. In total, the whole body is synergistically and automatically flexed. In the cases of jumping up and touching down, all the joints are coordinated by the complex muscular–skeletal system.

This coordination can be modified by changing the tension of the biarticular muscles. Figure 1.15 shows how the jumping behavior of the robot changes as the tension of muscle #9 changes. The robot is driven only by knee extensor #3, and other monoarticular muscles are not excited (not supplied with air). The tension of muscle #9 is controlled by changing the opening duration of the valve that supplies air with the muscle before the jumping procedure. The longer the duration, the tenser the muscle becomes. The top figure shows the jumping behavior when no air is supplied to muscle #9. If muscle #9 is not supplied with air, the ankle is not extended when the knee is extended, and the body leans forward. If muscle #9 is supplied with air for 50 ms, weak coordination between the knee and ankle is generated, but this is not sufficient to make the robot jump upwards (the middle figure). If muscle #9

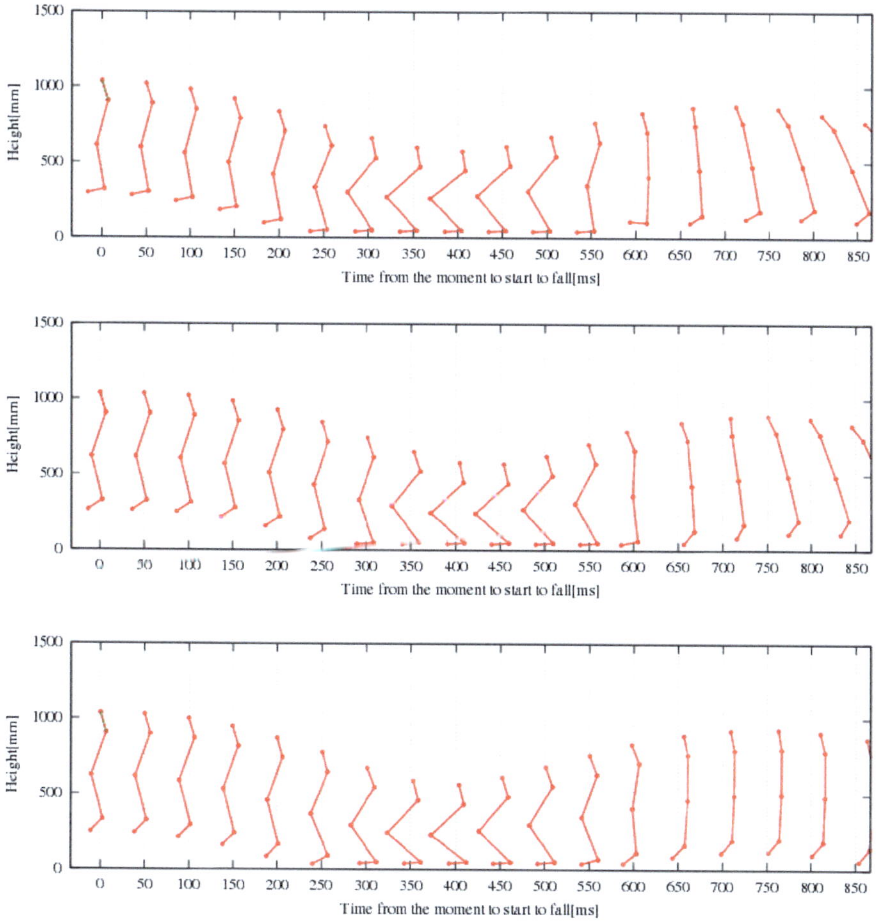

Fig. 1.15 Jumping behavior driven by knee extensor #3 (Hosoda et al. 2010). Changes in the jumping behavior of the robot as the tension of muscle #9 changes. If muscle #9 is totally relaxed (*top*), the robot leans forward. Weak coordination with the muscle is moderately tense (*middle*). Clear coordination between joints obtains when we sufficiently supply air to the muscle (*bottom*) (This figure and caption are modified from Hosoda et al. (2010), with kind permission from Springer Science and Business Media)

is fully supplied with air (the bottom figure), it coordinates the movement of the joints. The ankle joint is extended when the knee is extended, and as a result, the robot jumps straight up. This experimental result demonstrates that the robot can control the jumping direction by tuning the tension of muscle #9. This is, again, a very different way to control the jumping direction from the control of traditional motor-driven robots: when they want to change the jumping direction, they change the desired trajectory to track.

1.11 Summary and Conclusions

In this chapter, we have investigated the meaning of having a compliant body for the purpose of making robots achieve intelligent behavior. First, we have looked at a CCV and a glider and explained the trade-off between stability and controllability. The trade-off is relevant to a motor-driven robot and a passive dynamic walker. Then, we have looked at the design of an anthropomorphic robot arm that has a muscular–skeletal structure similar to that of humans, as well as its capabilities in achieving dynamic motion, sensing exploiting compliance, and dexterous manipulation. We have also looked at the muscular–skeletal legs that are designed on the basis of a passive dynamic walker, to understand their abilities in dynamic walking and sensing the terrain. Finally, we have increased the complexity of the legged robot to emulate the muscular–skeletal function of a human. These examples elucidate the dynamic and dexterous motions and sensing of muscular–skeletal robots. In other words, we have investigated *intelligence* in the structure of the human body and tried to reproduce it in muscular–skeletal robots. Obviously, the key idea is compliance. Compliance is the source of intelligence.

However, just being compliant is not enough. We must consider structured compliance, typically, compliance of a human body. As observed in many robot examples, reproducing a body comparable to that of a human helps us to have a better understanding of our embodied intelligence. Working from the human body and using it as a model help us to plan new types of robots with embodied intelligence comparable to human intelligence.

Exercises

Think about the following problems:

1. Find and discuss on a study demonstrating that an animal or insect exploits its dynamics for intelligent behavior. The reader might find some studies easily, but there are examples of dogs, cats, birds, bees, flies, and, of course, humans.
2. Find our daily tools/devices that are designed for our comfortable use, and exploit our body dynamics, and discuss how their designs work.

References

Braun, D.J., et al.: Robots driven by compliant actuators: optimal control under actuation constraints. IEEE Trans. Robot. **29**(5), 1085–1101 (2013)

Chou, C.P., Hannaford, B.: Measurement and modeling of McKibben pneumatic artificial muscles. IEEE Trans. Robot. Autom. **12**(1), 90–102 (1996)

Coleman, M.J., Chatterjee, A., Ruina, A.: Motions of a rimless spoked wheel: a simple 3d system with impacts. Dynam. Stabil. Syst. **12**(3), 139–160 (1997)

Hosoda, K., et al.: Pneumatic-driven jumping robot with anthropomorphic muscular skeleton structure. Auton. Robots **28**(3), 307–316 (2010)
Hosoda, K., et al.: Anthropomorphic muscular-skeletal robotic upper limb for understanding embodied intelligence. Adv. Robot. **26**(7), 729–744 (2012)
Jacobs, R., Bobbert, M.F., van Ingen Schenau, G.J.: Mechanical output from individual muscles during explosive leg extensions: the role of biarticular muscles. J. Biomech. **29**(4), 513–523 (1996)
McGeer, T.: Passive dynamic walking. Int. J. Robot. Res. **9**(2), 62–82 (1990)
Pfeifer, R., Bongard, J.: How the Body Shapes the Way We Think, Chapter 3. The MIT Press, Cambridge, MA (2007)
Takuma, T., Hosoda, K.: Terrain negotiation of a compliant biped robot driven by antagonistic artificial muscles. J. Robot. Mechatron. **19**(4), 423–428 (2007a)
Takuma, T., Hosoda, K.: Controlling walking behavior of passive dynamic walker utilizing passive joint compliance. In: Proceedings of 2007 IEEE/RSJ International Conference on Intelligent Robots and Systems, pp. 2975–2980 (2007b)
Wisse, M., van Frankenhuyzen, J.: Design and control of 'Mike'; A 2D autonomous biped based on passive dynamic walking. In: Proceedings of International Conference on Adaptive Motion of Animals and Machines (AMAM) (2003)
Yoshikawa, T.: Foundations of Robotics. The MIT Press, Boston (1990)

Chapter 2
Motor Control Based on the Muscle Synergy Hypothesis

Hiroaki Hirai, Hang Pham, Yohei Ariga, Kanna Uno, and Fumio Miyazaki

Abstract In neuroscience, the idea that motor behaviors are constructed by a combination of building blocks has been supported by a large amount of experimental evidences. The idea has been very attractive as a powerful strategy for solving the motor redundancy problem. While there are some candidates for motor primitives, such as submovements, oscillations, and mechanical impedances, synergies are one of the candidates for motor modules or composite units for motor control. Synergies are usually extracted by applying statistic techniques to explanatory variables, such as joint angles and electromyography (EMG) signals, and by decomposing these variables into fewer units. The results of factor decomposition are, however, not necessarily interpretable with these explanatory variables, even though the factors successfully reduce the dimensionality of movement; therefore, the physical meaning of synergies is unclear in most cases. To obtain insight into the meaning of synergies, this chapter proposes the agonist-antagonist muscle-pair (A-A) concept and uses other explanatory variables: the A-A ratio, which is related to the equilibrium point (EP), and the A-A sum, which is associated with mechanical stiffness. The A-A concept can be regarded as a form comparable to the EP hypothesis (EPH, λ model), and it can be extended to the novel concept of EP-based synergies. These explanatory variables enable us to identify muscle synergies from human EMG signals and to interpret the physical meaning of the extracted muscle synergies. This chapter introduces the EMG analysis in hand-force generation of a human upper limb and shows that the endpoint (hand) movement is governed by two muscle synergies for (1) radial movement generation and (2) angular movement generation in a polar coordinate system centered on the shoulder joint. On the basis of the analysis, a synergy-based framework of human motor control is hypothesized, and it can explain the mechanism of the movement control in a simple way.

H. Hirai (✉) • H. Pham • Y. Ariga • K. Uno • F. Miyazaki
Graduate School of Engineering Science, Osaka University, Toyonaka, Osaka, Japan
e-mail: hirai@me.es.osaka-u.ac.jp

M. Kasaki et al. (eds.), *Cognitive Neuroscience Robotics A*,
DOI 10.1007/978-4-431-54595-8_2

Keywords Human motor control • Muscle synergy • Equilibrium point (EP) • Joint stiffness • Agonist-antagonist muscle-pair (A-A) concept • A-A ratio • A-A sum • Pneumatic artificial muscle (PAM) • Electromyography (EMG) • Principal component analysis (PCA) • Hand force

2.1 Introduction

The flexibility of human movement is governed by the redundant degrees-of-freedom problem posed by Bernstein (1967). The movements we make, even the simplest ones, are the results of the coordinated actions of multiple muscles and joints of the limbs and trunk. The redundancy of degrees of freedom caused by multiple muscles and joints produces numerous possible solutions to a given task. Moreover, similar joint trajectories can be created by different muscle activation patterns (Gottlieb 1998). This biomechanical redundancy requires the central nervous system (CNS) to solve the problem of choosing a unique solution for controlling task variables. Unfortunately, the physiological mechanism of how the brain controls muscles still remains unknown. There are several hypotheses that purport to explain how the CNS uniquely specifies control commands to each muscle and to interpret the nature of the physiological control of the muscles by the brain. The synergy hypothesis suggests that muscle synergies, a specific pattern of muscle coactivation, provide a solution to the redundancy problem (Bizzi et al. 2002; d'Avella et al. 2003). Bernstein first proposed that the CNS might simplify the control of movement by coupling small groups of muscles into more global synergic units, thus reducing the number of controlled variables (Bernstein 1967). Many researchers have experimentally supported the idea that the CNS can generate a wide range of behavioral movements by combining groups of muscles or muscle synergies (Bizzi et al. 2008; d'Avella and Bizzi 2005; Giszter et al. 2007). In addition, Ting used muscle synergies to explain the mechanism of the transformation of neural commands into specific muscle activation patterns related to task-level variables (Ting 2007). Muscle synergies transform the desired control of task variables into high-dimensional muscle activations to produce biomechanical outputs that generate sensory signals mapping onto task variables. This framework provides the hierarchal control that the CNS uses to regulate muscle synergies which reflect task-variable information. However, the redundant degrees-of-freedom problem also exists in the transformation between muscle activation patterns and kinematic patterns of movement. This additional redundancy arises not only because multiple muscles cross each joint but also because the biomechanical equations of motion are such that different temporal patterns of muscle activation can lead to similar joint trajectories (Gottlieb 1998; Ting 2007). To solve this problem, it is necessary to reconsider the meaning of motor commands the CNS sends to each muscle. A promising hypothesis to interpret the physiology of movements is the equilibrium-point (EP) hypothesis (EPH, λ model) proposed by Feldman (1966). According to this hypothesis, the CNS sends a set of motor commands, each consisting of a reciprocal command and a coactivation command, to peripheral muscles, to select a desired EP and its apparent stiffness. The reciprocal command influences the change in the EP. The coactivation command influences the change in stiffness. There have been several attempts to reconstruct these motor commands during natural actions. All these studies, however, have produced questionable results because observed variables, such as forces and displacements, are indirect reflections of the motor commands (Latash 2008a). Other approaches to identifying the motor commands have been built on analysis of electromyographic (EMG) patterns of muscles. One approach is based on the concepts of the agonist-antagonist muscle-pair ratio (A-A ratio) and the agonist-antagonist muscle-pair sum (A-A sum). Note that the A-A ratio is different from the reciprocal command described by Feldman (1966). The derivation of the A-A concept is based on the analogy between the biological system and a robot system with antagonistic pneumatic artificial muscles (PAMs). For biological muscles, the equilibrium length and stiffness of a muscle can be changed by muscle activation (Shin et al. 2009). Analogously, the natural length and elastic coefficient of an artificial muscle vary with the internal air pressure. The A-A ratio for a PAM system, defined as the ratio of the air pressure of the extensor artificial muscle and the sum of the air pressure of the extensor and flexor artificial muscles, is directly and linearly related to the EP for a desired motion (Ariga et al. 2012a,b). The A-A sum for a PAM system, defined as the sum of the air pressure of the extensor and flexor artificial muscles, is directly and linearly associated with its mechanical stiffness at any EP (Ariga et al. 2012a). It is expected that these concepts would be effectively applicable to the biological system, especially in extracting muscle synergies from human EMG signals.

In what follows, we first describe the analogy between the biological system and a robot system with antagonistic PAMs. The analogy leads to a motor control scheme based on the A-A concept comparable to but distinct from the EPH. This implies that we can extract the kinematic command and stiffness command from EMG signals by using the A-A ratio and A-A sum. Next, we demonstrate how to extract muscle synergies from EMG signals by applying a dimensionality-reduction technique (e.g., principal component analysis (PCA)) to the datasets of the A-A ratio or A-A sum. Lastly, we present a framework for understanding how the CNS uniquely specifies control commands to each muscle.

2.2 Analogy Between Biological System and Robotic System with Antagonistic PAMs

This section explains an analogy between the biological system and a robotic system with antagonistic artificial muscles. First, theoretical models for a single-joint robotic system driven by two PAMs are conducted by introducing the key concept of A-A ratio and A-A sum. Then, under the A-A concept, a novel control method for controlling the equilibrium-joint angle and joint stiffness of the system is proposed, and the relationship between the A-A concept and an influential motor control hypothesis of EPH is discussed. Furthermore, the proposed A-A concept is extended to motor control for multi-joint movement.

2.2.1 Single PAM Model

The McKibben PAM is an actuator first invented in the 1950s by McKibben to motorize pneumatic arm orthosis (Tondu and Lopez 2000). This unique actuator is capable of converting the pneumatic energy into mechanical work for yielding a force, displacement, and mechanical impedance. The robotic system with this soft actuator has been studied since the 1950s, and recently the flexibility of the antagonistic system with PAMs is focused on with the analogy with the biological system. It is, however, difficult to realize accurate control by using the PAM. The nonlinear characteristics of PAM (e.g., friction, viscous elasticity, and hysteresis) are major obstacles in constructing a PAM system. To avoid these obstacles, Fujimoto et al. proposed a linear approximation model of PAM (Fujimoto et al. 2007). They modified the general PAM model originally proposed by Chou and Hannaford (1996) and presented a detailed PAM model, to explain the energy loss caused by friction and elasticity. An advantage of the Fujimoto model is that the characteristics of a PAM are expressed by a linear equation of contraction force and a contraction ratio, while the model considers complex PAM characteristics, such as mechanical energy loss. In this section, we describe the explicit model of a PAM based on the Fujimoto model. In the following equations, PPa is the internal air pressure of PAM, Vm^3 is the volume of PAM, FN is the force produced by PAM, lm is the length of PAM, and F_{diss}N is the energy dissipation of PAM due to the elastic force and movement friction force of rubber and sleeve. The equation is then obtained from the input-output relation of energy:

$$PdV = -(F + F_{\mathrm{diss}})dl \tag{2.1}$$

$$F_{\mathrm{diss}} = \alpha L_0 \varepsilon + f \tag{2.2}$$

where αN/m is the loss coefficient of elastic force and fN is loss of dynamic friction. The contraction rate ε is defined by the following equation with the natural length L_0m of PAM:

$$\varepsilon = \frac{L_0 - l}{L_0} \tag{2.3}$$

In addition, it is known that the volume V can be approximated by using certain constant coefficients η_i ($i = 0,1,2$) (Kagawa et al. 1993):

$$V = \eta_2 \varepsilon^2 + \eta_1 \varepsilon + \eta_0 \tag{2.4}$$

From (2.3) and (2.4), the finite change in the volume V and the length l are given by

$$dV = (2\eta_2\varepsilon + \eta_1)d\varepsilon \tag{2.5}$$

$$dl = -L_0 d\varepsilon. \tag{2.6}$$

By substituting (2.2), (2.5), and (2.6) into (2.1), the following linear form of the contraction force F is obtained:

$$F = P(a_1\varepsilon + a_0) - (b_1\varepsilon + b_0) \tag{2.7}$$

where $a_1 = \frac{2\eta_2}{L_0}$, $a_0 = \frac{\eta_1}{L_0}$, $b_1 = \alpha L_0$, and $b_0 = f$. To focus on the muscle length l instead of ε, (2.7) could also be rewritten as

$$F = K(P)(l - l_0(P)) \tag{2.8}$$

where

$$K(P) = -\frac{a_1 P - b_1}{L_0} \tag{2.9}$$

$$l_0(P) = (1 - \varepsilon_0)L_0 = \frac{C_1}{K(P)} + C_2 \tag{2.10}$$

$$\varepsilon_0 = -\frac{a_0 P - b_0}{a_1 P - b_1} \tag{2.11}$$

$$C_1 = \frac{-a_0 b_1 + a_1 b_0}{a_1} \tag{2.12}$$

$$C_2 = \frac{a_0 + a_1}{a_1} L_0. \tag{2.13}$$

Equation (2.8) indicates the system whose elastic coefficient $K(P)$ and natural length $l_0(P)$ change according to the internal air pressure P.

This characteristic of a PAM is analogous to a characteristic of biological muscle. Figure 2.1 presents a schematic drawing of biological muscle characteristics. The characteristic curve of muscle shifts depends on the level of electrical stimulation e. Feldman et al. report from their experiments with cats that:

1. The natural length $l_0(e)$ of biological muscle distinguishes the shift amount of the characteristic curve.
2. The shift amount of the characteristic curve is related to the electrical signal e (motor command) from CNS.
3. The slope of the characteristic curve gradually increases with electrical stimulation changes.

The EPH, one of the most influential hypotheses of motor control, is based on these findings (Feldman and Levin 2008). The relationship between the contraction

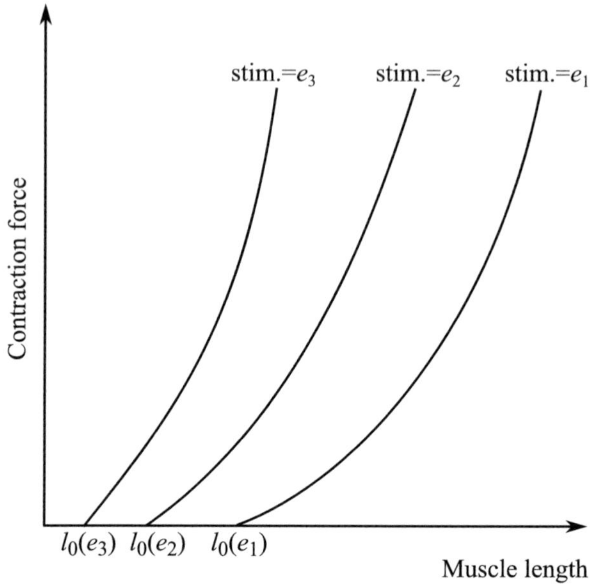

Fig. 2.1 Length-force properties of a biological muscle. $l_0(e_i)(i = 1, 2, 3)$ is the natural length of muscle, and e_i is the level of electrical stimulation ($e_1 < e_2 < e_3$). The slope of the curve indicates the elasticity (stiffness) of muscle

force and the muscle length of a biological muscle is clearly nonlinear under constant stimulation, but it may be approximated by the linear form in a limited range. Figure 2.2 shows the characteristics of a PAM. The straight lines in the figure indicate linear approximations calculated from the experimental data. According to (2.8), if the internal pressure P stays constant, the relationship between the contraction force F and the length l is linear. These length-force properties of PAM also show the following characteristics similar to characteristics of biological muscle:

1. The natural length $l_0(P)$ of PAM distinguishes the shift amount of the characteristic curve.
2. The shift amount of the characteristic curve is associated with the pneumatic pressure P (motor command) from the controller.
3. The slope of the curve $K(P)$ gradually increases according to the increase of the pneumatic pressure P.

These characteristics are formulated by (2.8), (2.9), and (2.10).

2.2.2 *Modeling the Agonist-Antagonist System with Two PAMs*

In a biological system, a joint movement is controlled by antagonistic pairings of muscles like the triceps and biceps. To explain our novel control method using the

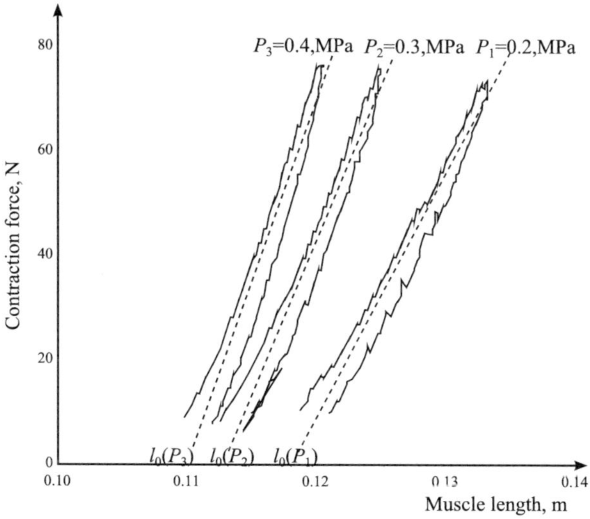

Fig. 2.2 Length-force properties of a PAM. The properties (including the natural length and elasticity (stiffness) of a PAM) are formulated by (2.8), (2.9), and (2.10). These characteristics are similar to those of a biological muscle

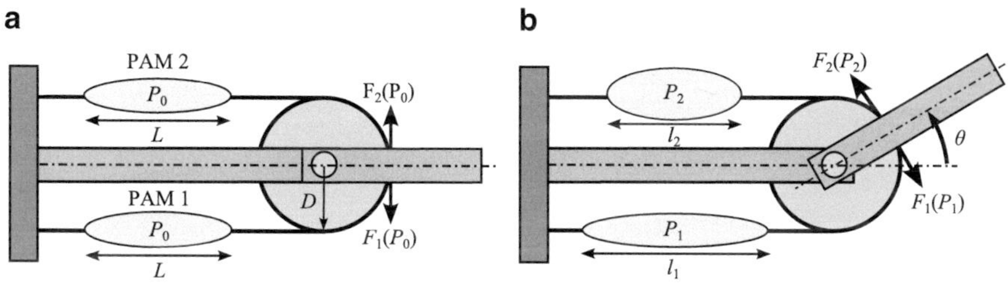

Fig. 2.3 Model of single-joint agonist-antagonist system with two PAMs: (**a**) initial state and (**b**) equilibrium state. For simplicity, it is assumed that the moment arm of a joint is constant and that the characteristics of two PAMs are the same

A-A concept, we start from considering a simple example of an agonist-antagonist system. Figure 2.3 shows the model of a single-joint arm with one pair of PAMs (agonist/antagonist). Let P_1Pa be the internal pressure of PAM1, P_2Pa be the internal pressure of PAM2, l_1m be the length of PAM1, l_2m be the length of PAM2, and θrad be the joint angle. Dm is the radius of the joint. The characteristics of the two PAMs are assumed to be the same. At the initial state, set $P_1 = P_2 = P_0$ and $l_1 = l_2 = L$.

2.2.2.1 EP and A-A Ratio

Consider the equilibrium-joint angle of the agonist-antagonist system with two PAMs. Let F_1N be the contraction force of PAM1 and F_2N be the contraction force of PAM2. The contraction forces of the two PAMs balance each other in the equilibrium state. Substituting (2.8) into the equation of $F_1 = F_2$ yields

$$K(P_1)(l_1 - l_0(P_1)) = K(P_2)(l_2 - l_0(P_2)). \tag{2.14}$$

Using the geometry constraint $D \cdot \theta = l_1 - L = L - l_2$ (Fig. 2.3),

$$\theta = \frac{C_2 - L}{D} \cdot \frac{K(P_1) - K(P_2)}{K(P_1) + K(P_2)}. \tag{2.15}$$

$K(P_i)$ ($i = 1, 2$) can also be rewritten as

$$K(P_i^*) = -\frac{a_1}{L_0} P_i^* = kP_i^* \tag{2.16}$$

by referring to (2.9) and defining $P_i^* = P_i - c$, where $k = -\frac{a_1}{L_0}$ and $c = \frac{b_1}{a_1}$. Using (2.15) and (2.16), θ is then given by

$$\theta = \frac{C_2 - L}{D} \cdot \frac{P_1^* - P_2^*}{P_1^* + P_2^*} \tag{2.17}$$

$$= \frac{2(C_2 - L)}{D} \cdot \left(\frac{P_1^*}{P_1^* + P_2^*} - \frac{1}{2} \right). \tag{2.18}$$

Furthermore, if the A-A ratio R is defined as

$$R = \frac{P_1^*}{P_1^* + P_2^*}, \tag{2.19}$$

then the relationship between θ and R is expressed by a simple equation. By substituting (2.19) into (2.18), the linear relationship between θ and R is finally obtained as

$$\theta = M(R - \frac{1}{2}) \tag{2.20}$$

where $M = \frac{2(C_2 - L)}{D}$. The variable M becomes constant when the radius of the joint D remains constant. The above expression has three advantages over other methods for controlling the agonist-antagonist system:

1. The A-A ratio R gives the relationship between the internal pressures of two PAMs (P_1, P_2).
2. Equation (2.20) provides the unique solution of the A-A ratio R according to the equilibrium-joint angle θ (or EP), although there are infinite candidates of PAM pressures that can achieve the EP.
3. The linear relationship between the EP θ and A-A ratio R enables us to control a joint movement in a simple way.

2.2.2.2 Joint Stiffness and A-A Sum

Next, consider the joint stiffness of the agonist-antagonist system with two PAMs (Fig. 2.3). The finite change in the joint angle from EP θ to $\theta + \Delta\theta$ causes the restoring force driving the joint angle toward EP. Let ΔF_1 and ΔF_2 be the generated forces by PAM1 and PAM2. Two agonist and antagonist forces are given by

$$\Delta F_1 = K(P_1)D\Delta\theta \tag{2.21}$$

$$\Delta F_2 = -K(P_2)D\Delta\theta. \tag{2.22}$$

Since the radius of the joint is D, the restoring torque $\Delta\tau$Nm is written as follows:

$$\begin{aligned} \Delta\tau &= (\Delta F_2 - \Delta F_1)D \\ &= -(K(P_1) + K(P_2))D^2\Delta\theta. \end{aligned} \tag{2.23}$$

If the A-A sum is defined as

$$S = P_1^* + P_2^*, \tag{2.24}$$

then the joint stiffness $G(= |\frac{\Delta\tau}{\Delta\theta}|)$Nm/rad can be written using (2.16), (2.23), and (2.24):

$$G = kD^2(P_1^* + P_2^*) = N \cdot S \tag{2.25}$$

where $N = kD^2$. The variable N is constant when the radius of the joint D stays constant. The advantages of this expression are the following:

1. Using the A-A sum S, we can uniquely set the joint stiffness G at any EPs.
2. The linear relationship between the stiffness G and the A-A sum S simplifies the design of the control system.

Thus, the control method with the A-A ratio R and the A-A sum S enables us to easily and separately control the equilibrium-joint angle and joint stiffness for a single-joint agonist-antagonist system with two PAMs. These explicit expressions are summarized by (2.20) and (2.25).

2.2.2.3 Relationship Between A-A Concept and EPH

EPH (λ model) is one of the promising hypotheses about motor control (Feldman 1966; Feldman and Levin 2008; Feldman et al. 1990). This control theory has survived from severe criticisms since Feldman proposed it in the mid-1960s. A basic concept of EPH is presented in Fig. 2.4. Each curve represents the relationship between joint angle and torque produced by a muscle. An “extensor” grows with decreasing joint angle (flexion side), and a “flexor” grows with increasing joint

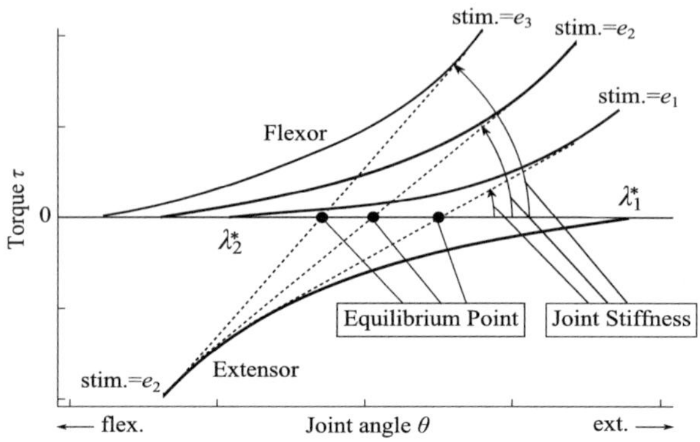

Fig. 2.4 Motor control based on EPH (λ model). EP is the position where joint torques by two antagonistic muscles balance each other. The slope of a *dashed line* indicates joint stiffness at each EP. EP and joint stiffness are controlled with the natural lengths of muscles (λ_1^*, λ_2^*) which are modulated by CNS. The reciprocal command $\frac{\lambda_1^*+\lambda_2^*}{2}$ affects EP, whereas the coactivation command $\frac{\lambda_1^*-\lambda_2^*}{2}$ affects joint stiffness

Fig. 2.5 Motor control based on the A-A concept. EP is the position where joint torques by two PAMs balance each other. The slope of a *dashed line* indicates joint stiffness at each EP. The A-A ratio R and the A-A sum S directly specify EP and joint stiffness through linear equations described by (2.20) and (2.25)

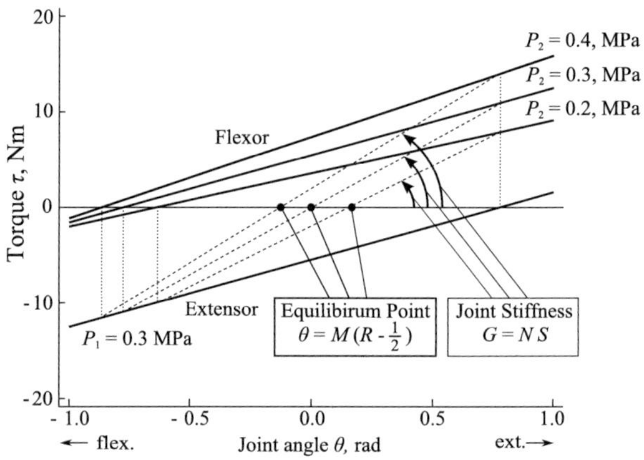

angle (extension side). EP is the position where two torques by the extensor and flexor balance each other. According to EPH, the natural lengths of muscles (λ_1^*, λ_2^*) are modulated by CNS, with the reciprocal command $\frac{\lambda_1^*+\lambda_2^*}{2}$ affecting EP and the coactivation command $\frac{\lambda_1^*-\lambda_2^*}{2}$ affecting joint stiffness. Figure 2.5 illustrates the relationship between EP and the joint stiffness of the agonist-antagonist system with two PAMs from the viewpoint of the A-A concept. (The graph is plotted using the experimental data.) An "extensor" grows with decreasing joint angle (flexion side), and a "flexor" grows with increasing joint angle (extension side). The dashed line represents the joint torque that the system produces, and EP is the position where the joint torque is 0 Nm/rad (i.e., two forces by two PAMs balance each other.). Joint stiffness is represented as the slope of the dashed line. The A-A concept is similar

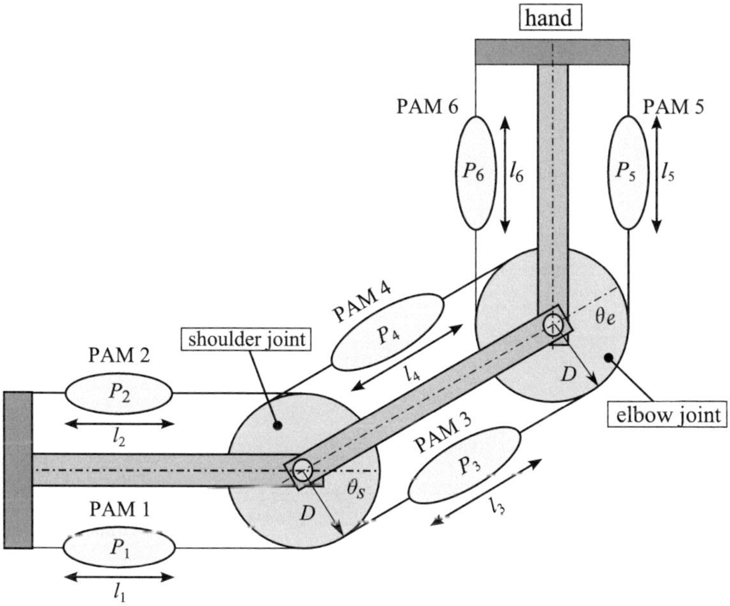

Fig. 2.6 Two-joint model with three antagonistic pairs of PAMs. PAMs 1 and 2, PAMs 3 and 4, and PAMs 5 and 6 are paired, respectively. For simplicity, it is assumed that the moment arm of each joint is constant and that the characteristics of PAMs are the same. This model replicates an antagonistic structure with multiple muscles in a human upper limb

to EPH in focusing on EP control; however, when we use the A-A ratio R and A-A sum S in the agonist-antagonist system, the EP θ is represented linearly using the A-A ratio R and the joint stiffness G is also represented linearly using the A-A sum S (see (2.20) and (2.25)). These parameters can be controlled individually for a single-joint agonist-antagonist system.

2.2.2.4 Extension of the A-A Concept to Motor Control of Two-Joint System with Multiple PAMs

Consider a two-joint PAM model that mimics a human arm structure with shoulder and elbow joints and three antagonistic pairs of muscles around and connecting the two joints. The structure of the PAM model and its parameters are illustrated in Fig. 2.6. We assume the initial state at which the shoulder-joint angle is $\theta_s = 0$, the elbow-joint angle is $\theta_e = 0$, and the internal pressure and length of PAM are $P_i = P_0$ and $l_i = L$ $(i = 1, 2, \ldots, 6)$, respectively. When θ_s and θ_e are at the equilibrium state, the contraction forces F_i of PAMs i balance each other, such that

$$F_1 - F_2 + F_3 - F_4 = 0 \tag{2.26}$$

$$F_3 - F_4 + F_5 - F_6 = 0. \tag{2.27}$$

Considering (2.8), (2.26) and (2.27) can be rewritten as

$$K(P_1)(l_1 - l_0(P_1)) - K(P_2)(l_2 - l_0(P_2)) + K(P_3)(l_3 - l_0(P_3)) - K(P_4)(l_4 - l_0(P_4)) = 0 \quad (2.28)$$

$$K(P_3)(l_3 - l_0(P_3)) - K(P_4)(l_4 - l_0(P_4)) + K(P_5)(l_5 - l_0(P_5)) - K(P_6)(l_6 - l_0(P_6)) = 0. \quad (2.29)$$

The geometry constraint condition brings

$$l_1 = L + D\theta_s \quad (2.30)$$

$$l_2 = L - D\theta_s \quad (2.31)$$

$$l_3 = L + D(\theta_s + \theta_e) \quad (2.32)$$

$$l_4 = L - D(\theta_s + \theta_e) \quad (2.33)$$

$$l_5 = L + D\theta_e \quad (2.34)$$

$$l_6 = L - D\theta_e \quad (2.35)$$

where D is the moment arm of the joints. Given (2.10), substituting above equations from (2.30) to (2.35) into (2.28) and (2.29), the following can be obtained:

$$\begin{bmatrix} K(P_1) + K(P_2) + K(P_3) + K(P_4) & K(P_3) + K(P_4) \\ K(P_3) + K(P_4) & K(P_3) + K(P_4) + K(P_5) + K(P_6) \end{bmatrix} \begin{bmatrix} \theta_s \\ \theta_e \end{bmatrix} = \frac{C_2 - L}{D} \begin{bmatrix} K(P_1) - K(P_2) + K(P_3) - K(P_4) \\ K(P_5) - K(P_6) + K(P_3) - K(P_4) \end{bmatrix}. \quad (2.36)$$

By constructing $P_i^* = P_i - c$, $K(P_i)$ can be rewritten as $K(P_i^*) = kP_i^*$ (see (2.16)). Then, by defining the A-A ratio and A-A sum as

$$R_i = \frac{P_{2i-1}^*}{P_{2i-1}^* + P_{2i}^*} \quad (2.37)$$

$$S_i = P_{2i-1}^* + P_{2i}^*, \quad (2.38)$$

the equilibrium-joint angles of θ_s and θ_e in (2.36) can be rewritten as follows:

$$\boldsymbol{\theta} = \boldsymbol{MS}(\boldsymbol{R} - \frac{1}{2}\boldsymbol{U}) \quad (2.39)$$

where $\boldsymbol{\theta} = [\theta_s, \theta_e]^T$, $\boldsymbol{M} = \frac{2(C_2-L)}{D}$, $\boldsymbol{R} = [R_1, R_2, R_3]^T$, $\boldsymbol{U} = [1, 1, 1]^T$ and

$$\boldsymbol{S} = \begin{bmatrix} \frac{S_1S_2+S_1S_3}{S_1S_2+S_2S_3+S_1S_3} & \frac{S_2S_3}{S_1S_2+S_2S_3+S_1S_3} & \frac{-S_2S_3}{S_1S_2+S_2S_3+S_1S_3} \\ \frac{-S_1S_2}{S_1S_2+S_2S_3+S_1S_3} & \frac{S_1S_2}{S_1S_2+S_2S_3+S_1S_3} & \frac{S_1S_3+S_2S_3}{S_1S_2+S_2S_3+S_1S_3} \end{bmatrix}. \quad (2.40)$$

Note that (2.39) is an advanced notation of (2.20). Equation (2.39) indicates that if the matrix $\boldsymbol{S}$ is constant owing to the superb balance among S_1, S_2, and S_3, the relationship between $\boldsymbol{\theta}$ and $\boldsymbol{R}$ is linear. This leads to a simple way to control the system that replicates the human arm structure. The simple control can be realized, for example, when $\frac{S_1}{S_2}$ and $\frac{S_3}{S_2}$ are constant.

2.3 Muscle Synergy Hypothesis in Human Motor Control

This section presents the analysis of the hand-force generation of a human upper limb. The investigation into how muscle activities generate hand forces under an isometric condition gives a clue for understanding how we control the muscle groups in our limb to achieve a desired movement. Using the A-A concepts proposed in the previous section, we can investigate the physical meaning of muscle coordination during a task. This section also presents a novel framework for the motor control of upper limb movement on the basis of the statistic analysis of EMG activities under the A-A concept.

2.3.1 Materials and Methods

2.3.1.1 Experimental Protocol

The subjects of the experiment are three healthy volunteers (all of whom are 22 years old, male, and right handed) with no record of neuromuscular deficits. The experimental procedures were conducted with the approval of the ethics committee at Osaka University. The subjects were asked to produce 8N of hand force pointing along eight directions under an isometric condition in a horizontal plane, starting at direction 1 and shifting orientation by 45° in the counterclockwise rotation until ending at direction 8 (Fig. 2.7a). The hand position was 0.28 m in front of the subject's trunk. The task goal was to produce hand force as close to the reference force as possible and to keep maintaining the force in a 6-s duration before moving to the next direction. The task was performed with the left hand and then with the right hand, and it was identical for both hands. The subjects sat comfortably on chairs, with the elbow supported to reduce the gravitational effect and to allow for shoulder and elbow flexion-extension in a horizontal plane. The subjects grasped a joystick and pulled or pushed it to produce hand forces at a comfortable speed while looking at reference forces displayed on a screen. For simplicity, wrist movement was prevented by a splint so that it could be ignored in

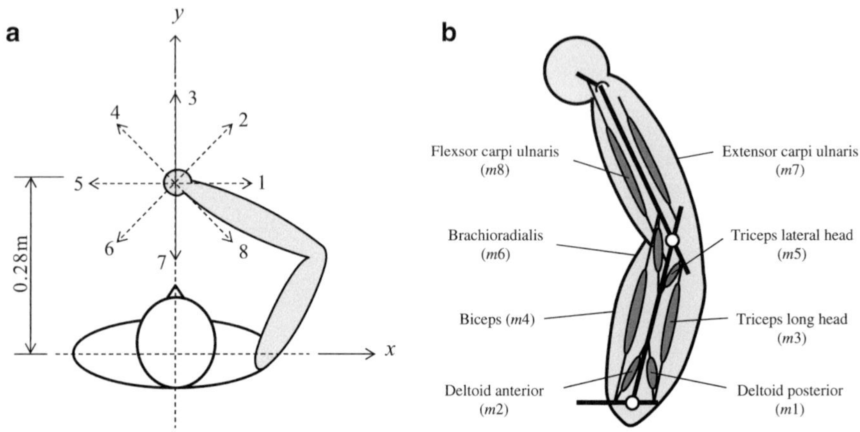

Fig. 2.7 (**a**) Experiment setup, top view. Subjects were asked to produce hand force along eight directions in order, from direction 1 to direction 8, under an isometric condition. The task was performed with a left/right arm, and it was identical for both arms. (**b**) Sketch of examined eight muscles which mainly contribute to the studied movement

the analysis. The EMGs of eight muscles (see Fig. 2.7b) that mainly contribute to the studied task were collected by using a multi-telemeter system (WEB-5000, Nihon Kohden Corp., Japan) at 1000 Hz. Examined muscles were identified according to the guidelines in Hislop and Montgomery (2007). After cleansing the skin to reduce the resistance below 10 kΩ, surface electrodes were placed on the examined muscles. EMG signals were band-pass filtered (0.03–450 Hz), hum filtered (60 Hz), amplified (×2000), and stored in a computer. A force sensor (USL06-H5-200, Tech Gihan Co., Ltd., Japan) was attached under the joystick to measure hand forces. Initially, we measured the maximum voluntary contraction (MVC) or the maximum value of the EMGs of the subjects' examined muscles. The subjects then practiced the task several times and became familiar with the system and the task goal. Data of these practices were not stored. After practicing, the subjects performed the experiment trials. For every trial, the EMGs and hand forces were synchronized and collected at a sampling rate of 1000 Hz.

2.3.1.2 Data Analysis

(1) EMG preprocessing

The raw EMG signals, which contain a significant amount of noise and artifacts, need to be preprocessed to be more reliable for the analysis. The preprocess of EMGs includes the following steps: (1) Band-pass filter (10–150 Hz) the raw EMGs to reduce anti-aliasing effects within sampling. (2) Full-wave rectify the signals to make them more readable. (3) Smooth the signals by a moving average window that brings out the mean trend of EMG development. (4) Amplitude normalize to

MVC to eliminate the influence of the detection condition and to make the data comparable between different muscles and different subjects.

(2) Data normalizing

The dataset of EMGs in each trial is a $n \times 8$ matrix:

$$\boldsymbol{m}(t) = \begin{bmatrix} m_1(t_1) & \cdots & m_j(t_1) & \cdots & m_8(t_1) \\ \vdots & \vdots & \vdots & \vdots & \vdots \\ m_1(t_n) & \cdots & m_j(t_n) & \cdots & m_8(t_n) \end{bmatrix} \tag{2.41}$$

where $m_j(t_n)$ is the time-varying EMG of j-th muscle at time t_n and n indicates the time points in a trial. The hand-force dataset also includes the same time points. To observe the whole process of producing hand forces in each trial, the EMGs and hand forces were averaged with respect to time. Moreover, the averaged EMGs were standardized as follows:

$$m_j^*(k) = \frac{m_j^k}{\sqrt{\sum_{k=1}^{8}(m_j^k - \overline{m}_j)^2}} \tag{2.42}$$

where m_j^k is the averaged EMG of the j-th muscle m_j at the k-th direction and $\overline{m}_j$ is the average of m_j^k over all directions. Here, a trial includes movements in eight directions, and thus $k = 1, \ldots, 8$. The dataset of the averaged EMGs in a trial is a 8×8 matrix corresponding to the directions:

$$\boldsymbol{M}^* = \begin{bmatrix} m_1^*(1) & \cdots & m_j^*(1) & \cdots & m_8^*(1) \\ \vdots & \vdots & \vdots & \vdots & \vdots \\ m_1^*(k) & \cdots & m_j^*(k) & \cdots & m_8^*(k) \\ \vdots & \vdots & \vdots & \vdots & \vdots \\ m_1^*(8) & \cdots & m_j^*(8) & \cdots & m_8^*(8) \end{bmatrix} \tag{2.43}$$

(3) A-A ratio and A-A sum datasets

For the PAM system, we defined the A-A ratio and A-A sum with respect to the pressures supplied to each PAM as appearing in (2.19) and (2.24), respectively. To apply these definitions to the human muscles, we selected the antagonistic muscle pairs from the anatomical point of view and modified them with respect to the EMGs of each muscle. The A-A ratio and A-A sum are defined, respectively, by

$$r_j = \frac{m_{2j-1}^*}{m_{2j-1}^* + m_{2j}^*} \tag{2.44}$$

$$s_j = m_{2j-1}^* + m_{2j}^* \tag{2.45}$$

Table 2.1 Definition of A-A ratio and A-A sum

A-A ratio	Physical property	A-A sum	Physical property
$r_1 = m_1^*/(m_1^* + m_2^*)$	Shoulder extension	$s_1 = m_1^* + m_2^*$	Shoulder stiffness increase
$r_2 = m_3^*/(m_3^* + m_4^*)$	Shoulder & elbow extension	$s_2 = m_3^* + m_4^*$	Shoulder & elbow stiffness increase
$r_3 = m_5^*/(m_5^* + m_6^*)$	Elbow extension	$s_3 = m_5^* + m_6^*$	Elbow stiffness increase
$r_4 = m_7^*/(m_7^* + m_8^*)$	Elbow (& wrist) extension	$s_4 = m_7^* + m_8^*$	Elbow (& wrist) stiffness increase

where j indicates the j-th muscle pair (j = 1, 2, 3, 4). Here, we addressed the four pairs of antagonistic muscles that mainly contributed to the studied task. Physical properties of A-A ratio and A-A sum are given in Table 2.1. By defining the A-A ratio as in (2.44) and (2.45), we can explain the kinematics of joint movement. For example, the kinematics of the shoulder joint can be elaborated by the change in A-A ratio $r_1 = \frac{m_1^*}{m_1^* + m_2^*}$. When m_1^* increases and m_2^* decreases, the A-A ratio r_1 increases. Because the extensor muscle m_1^* works harder than the flexor muscle m_2^*, the shoulder joint will extend. Hence, a change in the A-A ratio r_1 will influence the kinematics of shoulder-joint movement. This mechanism can be elucidated by the relationship between A-A ratios and EPs (or equilibrium-joint angles) as it is described in Sect. 2.2.2.4. The relationship between the A-A ratio and the EP also supports the physical meaning of the proposed A-A ratio (Ariga et al. 2012a). That is, when the agonist and antagonist muscles contract unequally or the A-A ratio varies, joint movement will be generated, driving the system to a new EP. In addition, the linear relationship between the A-A ratio and the EP also offers a simple way to control joint movement. On the other hand, because the stiffness of a muscle is proportional to its level of activity (Milner et al. 1995), the stiffness of a joint is proportional to the sum of the activity level of antagonistic muscles across the joint. Therefore, by defining an A-A sum as the sum of the EMGs of muscles of an antagonistic muscle pair, we can explain the contribution of each muscle pair to joint stiffness. For example, when both muscles m_1^* and m_2^* contract or m_1^* and m_2^* increase, the A-A sum $s_1 = m_1^* + m_2^*$ will increase. Physiologically, the contraction of these two muscles results in an increase of shoulder-joint stiffness. Hence, the A-A sum s_1 influences shoulder-joint stiffness. This relationship between the A-A sum and joint stiffness was experimentally proved in Ariga et al. (2012a) by using the antagonistic system with PAMs. In short, using the proposed A-A ratio and A-A sum, we expected to easily control the equilibrium-joint angle and joint stiffness, which would consequently generate arm movements.

For this hand-force generation task, the dataset of A-A ratios corresponding to the EMGs in (2.43) is a 8×4 matrix:

$$
\boldsymbol{R}^* = \begin{bmatrix} r_1(1) & r_2(1) & r_3(1) & r_4(1) \\ \vdots & \vdots & \vdots & \vdots \\ r_1(k) & r_2(k) & r_3(k) & r_4(k) \\ \vdots & \vdots & \vdots & \vdots \\ r_1(8) & r_2(8) & r_3(8) & r_4(8) \end{bmatrix} \tag{2.46}
$$

$$
= [\boldsymbol{r}(1), \ldots, \boldsymbol{r}(k), \ldots, \boldsymbol{r}(8)]^T \tag{2.47}
$$

where $\boldsymbol{r}(k) = [r_1(k), r_2(k), r_3(k), r_4(k)]^T$ $(k = 1, 2, \ldots, 8)$.

(4) Data reducing

A widely used statistical technique for reducing data dimensionality is PCA (Jolliffe 1986). For example, Artemiadis and Kyriakopoulos (2010) successfully embedded a high-dimensional dataset of muscle activations and corresponding joint angles in two manifolds of fewer dimensions. This reduction has two advantages. First, as for joint angles, compressed composite units for motor control are suggested by using fewer variables to describe movement. Second, for analysis reasons, it is attractive to represent back the two manifolds into the high-dimensional space. Therefore, it is suggested that, by applying PCA to datasets of the A-A ratio and by using the dimensionally reduced independent variables, we can find a new representation for the motion of the end effector in the Cartesian space. The data can be represented by linear combinations as

$$
\Delta \boldsymbol{r}(k) = \sum_{j=1}^{4} w_j(k) \boldsymbol{q}_j \tag{2.48}
$$

where $\Delta \boldsymbol{r}(k) = \boldsymbol{r}(k) - \bar{\boldsymbol{r}}$, with $\bar{\boldsymbol{r}}$ being the averaged value vector of r_i $(i = 1, 2, 3, 4)$ in all directions, and $w_j(k)$ and $\boldsymbol{q}_j$ are the j-th PC score and the j-th PC vector, respectively.

2.3.2 *Muscle Synergy Interpretation*

2.3.2.1 PCA Results

Principal components (PCs) are found from the covariance matrix of the A-A ratio dataset $\boldsymbol{R}^*$. The number of retained PCs was chosen so as to preserve the most information of A-A ratios. With regard to the studied movement, the first two PCs of the A-A ratio have contributed over 90 % of the total variance of the A-A ratio data. Table 2.2 shows the percentage of variance accounted for (VAF) by the first two PCs.

Here, we present the analysis result of only one subject (subject A) as the obtained results were consistent across all of the subjects. Figure 2.8 exhibits the

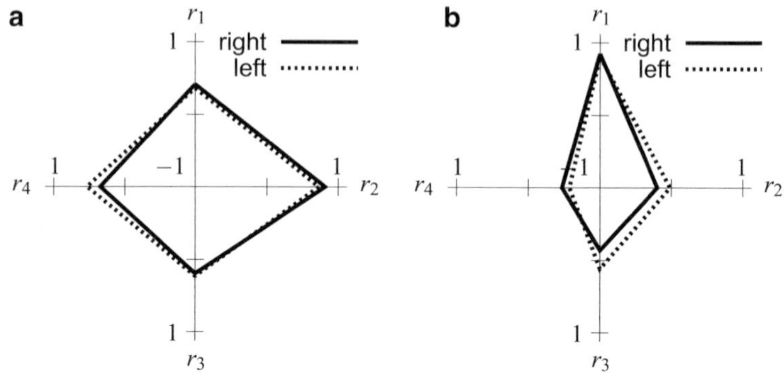

Fig. 2.8 PC vectors resulted from PCA: (**a**) PC1 vectors and (**b**) PC2 vectors. Each half of the axes indicates the value of each element corresponding to each A-A ratio of PC1 and PC2 vectors. The *solid* and *dashed lines* indicate the right-hand and the left-hand trials, respectively

Table 2.2 %VAF by the first two PCs

	Right hand			Left hand		
Subject	PC1	PC2	Total	PC1	PC2	Total
A	81.27	16.53	97.80	69.35	29.84	99.19
B	65.14	29.70	94.84	71.55	25.97	97.52
C	66.42	30.77	97.19	49.99	45.47	95.46

PC vectors. The left and right graphs represent PC1 and PC2 vectors, respectively. Each half of the axes indicates the value of the element corresponding to each A-A ratio of the PC vectors; the solid and dashed lines represent the right-hand and the left-hand trials, respectively. Each element of the PC vector is within the range of $[-1,1]$. The origin in each graph represents level -1. As indicated in the figure, the forms of the PC vectors of both hands are similar. All the elements of PC1 vectors are positive, and the element of PC2 vectors that corresponds to r_1 is positive; the element related to r_4 is negative, and the remaining two elements are negligible because they are close to zero. The even location of the elements of PC1 vectors on all the axes implies a similar contribution of all muscle extension, whereas the distribution of PC2 vector elements, weighting to r_1 and r_4, implies a simultaneous contribution of the shoulder extension and elbow flexion. PC vectors, therefore, can be regarded as representations of synergies which merge the coordination of multi-articular muscles. Particularly, in the case of hand-force generation, any force vector in a horizontal plane can be generated by two synergies. Muscle synergies, although usually difficult to understand from the original EMGs, become interpretable when they are represented by PC vectors derived from A-A ratio of EMGs. In other words, by using PC vectors, we can elucidate the physical meaning of muscle synergies more clearly. This characteristic will be discussed in detail in the next section. Figure 2.9 plots the scores of the first two PCs. The top and bottom rows present the observations of hand-force deviation (F_x, F_y) and PC scores (w_1, w_2), respectively.

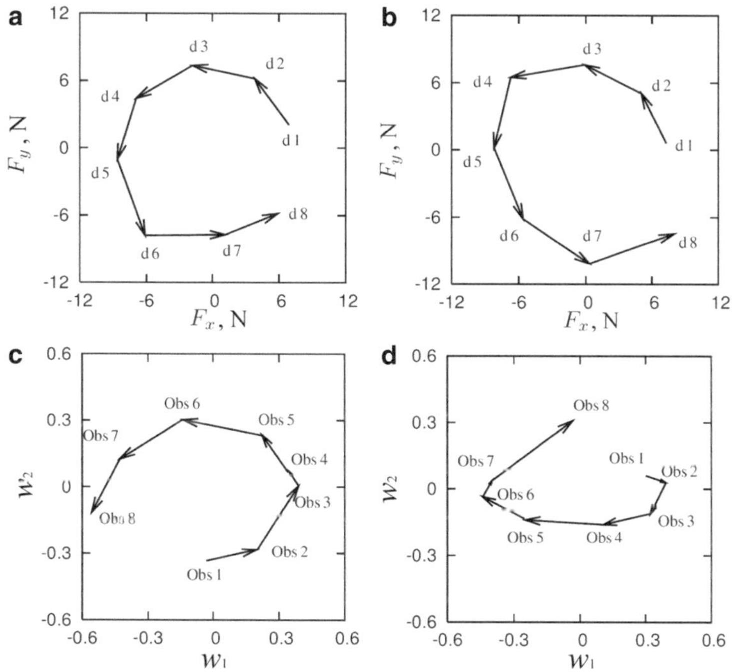

Fig. 2.9 Observation of hand force and PC scores corresponding to force vectors for the left-hand test and the right-hand test. (**a, b**) is hand-force observation in eight directions (d_ks, k=1,2,…,8); (**c, d**) is observation (Obs) of PC scores in eight directions. (**a, c**) shows the result of the left-hand test; (**b, d**) shows the result of the right-hand test

The left column is the result of the left-hand test, and the right column is the result of the right-hand test. As seen in this figure, the PC scores all go around a central point, just as the generated hand forces do. Note that the PC scores of the left hand are almost the same as those of the right hand at symmetric positions, reflecting the fact that the left and right arm postures are symmetric (Pham et al. 2011). Since the shape of the PC scores resembles the circle like shape of the hand force, a high correlation between PC scores and hand force is expected. This characteristic will be examined by analysis in the next section.

2.3.2.2 Hand-Force Estimation

For simplicity, we assume that hand force is generated only by manipulating the EPs (or the equilibrium-joint angles) defined by A-A ratios. Let $\Delta\boldsymbol{\theta}_{eq}$ be a deviation of the EP vector giving joint torques $\Delta\boldsymbol{\tau}$ as a result of joint stiffness such that

$$\Delta\boldsymbol{\tau} = \boldsymbol{K}\Delta\boldsymbol{\theta}_{eq} \tag{2.49}$$

where $\boldsymbol{K}$ is a joint stiffness matrix. The hand force can be converted into the equivalent joint torques according to the relationship $\Delta\boldsymbol{\tau} = \boldsymbol{J}^T\Delta\boldsymbol{F}$, where $\boldsymbol{J}^T$ is the transpose of the Jacobian matrix which represents the infinitesimal relationship between the joint displacements and the hand position. The hand force is, therefore, determined by the following linear equation:

$$\Delta\boldsymbol{F} = (\boldsymbol{J}^T)^{-1}\boldsymbol{K}\Delta\boldsymbol{\theta}_{eq}, \tag{2.50}$$

provided that the Jacobian is non-singular. Considering the linear relationship between the EP $\boldsymbol{\theta}_{eq}$ and the A-A ratio $\boldsymbol{R}$ (or $\boldsymbol{r}$) in (2.39), it is obtained by

$$\Delta\boldsymbol{F} = (\boldsymbol{J}^T)^{-1}\boldsymbol{KMS}\Delta\boldsymbol{r} \tag{2.51}$$

which means that the hand force can be controlled by the A-A ratio. It is worth noting that the A-A ratio data $\Delta\boldsymbol{r}(k)$ generates the hand force in the k direction and it can be approximately represented by the first two PCs of the A-A ratio as

$$\Delta\boldsymbol{r}(k) \fallingdotseq w_1(k)\boldsymbol{q}_1 + w_2(k)\boldsymbol{q}_2 \tag{2.52}$$

where $\boldsymbol{q}_1$ and $\boldsymbol{q}_2$ represent muscle synergies in terms of A-A ratio. The muscle synergies $\boldsymbol{q}_1$ and $\boldsymbol{q}_2$ are common for hand-force generation in any direction. Then (2.51) yields the following relationship between the hand force and PC vectors:

$$\Delta\boldsymbol{F}(k) \fallingdotseq w_1(k)\boldsymbol{A}\boldsymbol{q}_1 + w_2(k)\boldsymbol{A}\boldsymbol{q}_2 = w_1(k)\boldsymbol{p}_1 + w_2(k)\boldsymbol{p}_2 \tag{2.53}$$

where $\boldsymbol{A} = (\boldsymbol{J}^T)^{-1}\boldsymbol{KMS}$ is a linear mapping operator onto the hand-force vector space and $\boldsymbol{p}_1 = \boldsymbol{A}\boldsymbol{q}_1$ and $\boldsymbol{p}_2 = \boldsymbol{A}\boldsymbol{q}_2$ are muscle synergies represented in terms of hand-force vector. We henceforth call $(\boldsymbol{p}_1, \boldsymbol{p}_2)$ "synergy-force vectors." Assuming that the linear mapping operator $\boldsymbol{A}$ is constant, the synergy-force vectors can be estimated by the linear regression model of (2.53) with the hand-force data and corresponding PC scores shown in Fig. 2.9c, d. As indicated in Table 2.3, this model can explain approximately more than 90 % of the variance in the hand-force profiles (the regression coefficient, R^2, exceeds 90 % for both hands).

Figure 2.10 visually illustrates how well the predicted values fit the measured values. Figure 2.11 exhibits the normalized vectors of $(\boldsymbol{p}_1, \boldsymbol{p}_2)$ (denoted as $(\boldsymbol{p}_1^*, \boldsymbol{p}_2^*)$). As seen in this figure, in each hand, the two vectors $(\boldsymbol{p}_1^*, \boldsymbol{p}_2^*)$ are almost orthogonal

Table 2.3 Regression model for hand force by PC scores

	R^2 for F_x	R^2 for F_y	Regression model
Left-hand test	0.94	0.96	$\begin{bmatrix} F_x \\ F_y \end{bmatrix} = \begin{bmatrix} -8.49 \\ 14.88 \end{bmatrix} w_1 + \begin{bmatrix} -22.15 \\ -13.47 \end{bmatrix} w_2$
Right-hand test	0.91	0.95	$\begin{bmatrix} F_x \\ F_y \end{bmatrix} = \begin{bmatrix} 9.82 \\ 15.78 \end{bmatrix} w_1 + \begin{bmatrix} 34.4 \\ -26.73 \end{bmatrix} w_2$

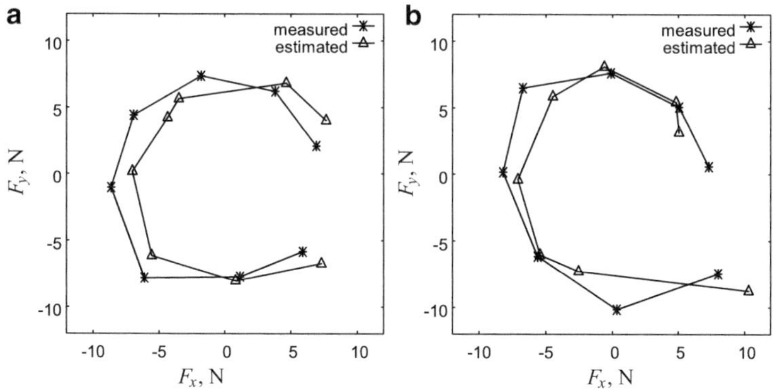

Fig. 2.10 Hand force and its estimation in (**a**) left-hand test and (**b**) right-hand test. The regression coefficient, R^2, exceeded 90 % for both hands

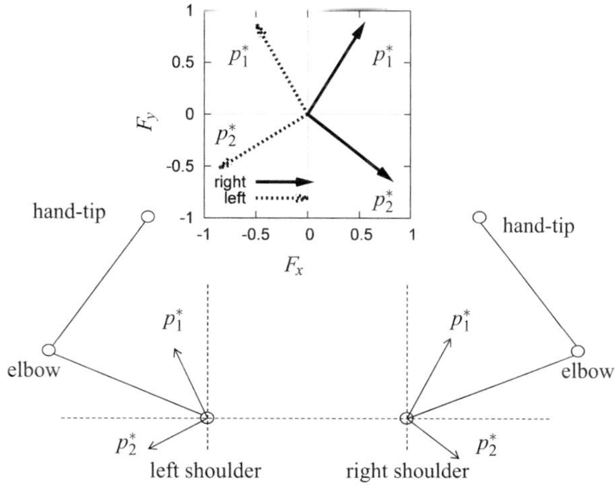

Fig. 2.11 Synergy-force vectors. The first synergy-force vector $\boldsymbol{p}_1^*$ tends to move forward to the endpoint (hand) position, and the second synergy-force vector $\boldsymbol{p}_2^*$ tends to cross to $\boldsymbol{p}_1^*$ at right angles. This indicates that the synergy-force vectors are represented as the bases in the polar coordinate system centered on the shoulder joint and implies that they may be motor modules for endpoint force control

to each other, and in both right and left hands, they are nearly symmetric. The physical meaning of this reflection is that, in respect to a polar coordinate frame centered on the shoulder joint, the first synergy represented by $\boldsymbol{p}_1$ seems to generate hand force in the radial direction and the second synergy represented by $\boldsymbol{p}_2$ seems to induce hand force in the angular direction. This result suggests that we can control movements by adjusting hand force or, more precisely, by changing the weight of $(w_1(k), w_2(k))$ to fit the direction and magnitude of the intended hand force. Hence, the result provides a simple method for dealing with the redundancy problem for the control of the musculoskeletal system. That is, using a few number of muscle synergies extracted from the A-A ratios of EMG signals, we can control pairs of antagonistic muscles and consequently activate limb movements. Along the same line, Ivanenko et al. (2004) applied PCA to EMGs collected from 12 to 16 muscles in a walking movement. They found that five component factors accounted

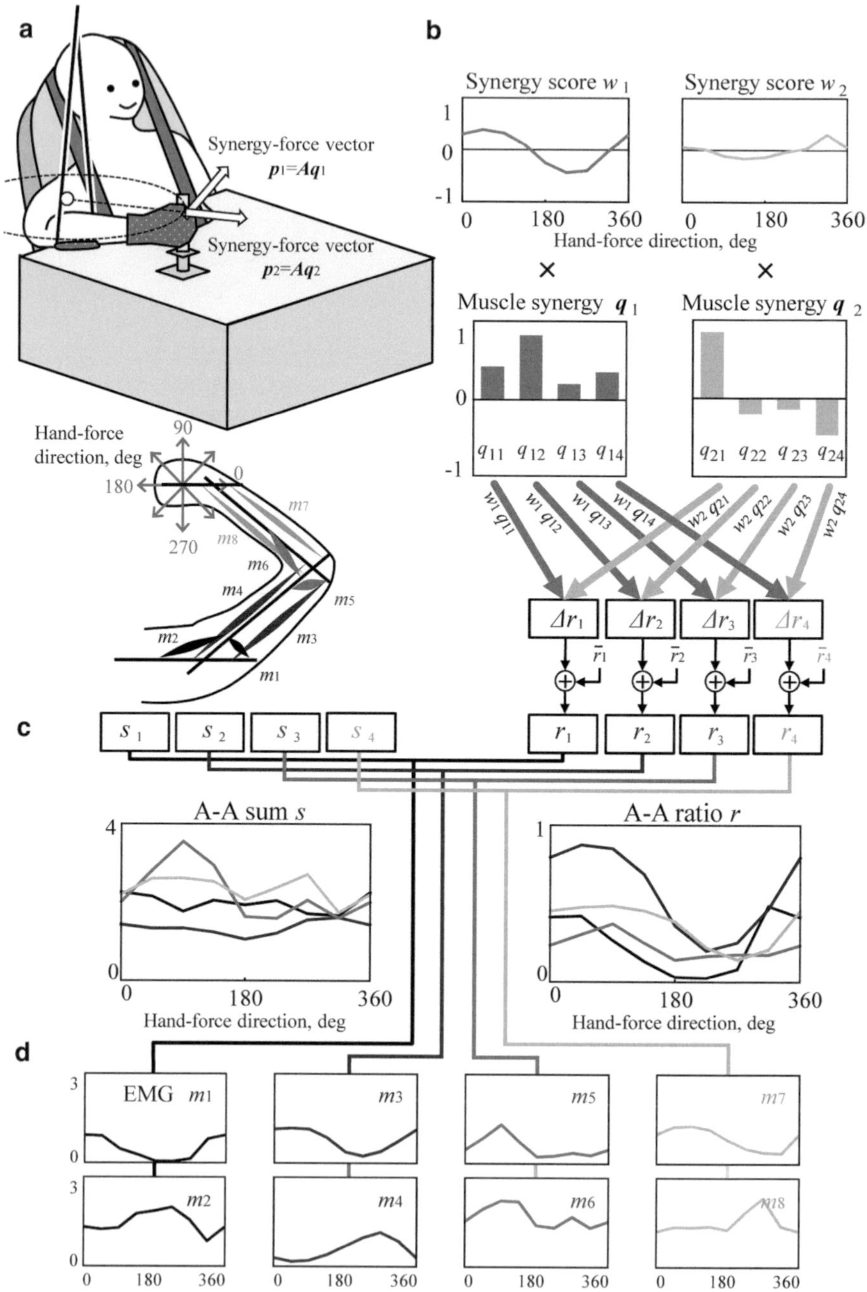

Fig. 2.12 Muscle synergy-based hierarchy model of human motor control. (**a**) Hand-force control task. A subject controls the endpoint force in eight directions under an isometric condition in a horizontal plane. (**b**) The A-A ratio $\boldsymbol{r}$ $(= [r_1, r_2, r_3, r_4]^T)$ is a set of motor commands generated by

for about 90 % of the total EMG patterns of activation muscles and showed that each component distinguishes the timing of muscle-group activation. They were, however, unable to interpret how these factors functionally grouped the muscles and how they were related to force demand during locomotion. In this study, we have applied PCA to the A-A ratio derived from EMGs. This analysis has two advantages. One advantage is the smaller dataset, and this makes the analysis more interesting. The other advantage is that, by using PCA algorithm on the dataset of the A-A ratio which governs the kinematics, the reduced dataset would reveal the relationship between muscle activations and hand-force production. Indeed, the regression analysis results imply a strong relationship between the two identified synergies that correspond to the two PCs resulting from PCA and hand-force generation.

2.3.3 A Novel Framework of Human Motor Control

On the basis of the results of PCA and the regression analysis on EMGs, we hypothesize a novel framework for the neuro-mechanical control of the human arm. The control scheme is shown in Fig. 2.12. This figure illustrates an example of using the framework to explain the hand-force task. The procedure includes: (1) The controller determines the task variables (hand forces $\boldsymbol{F}$). The task variables may include feedback elements. (2) The hand forces are converted into synergy variables (w_1,w_2) reflecting the synergy-force vectors ($\boldsymbol{p}_1$, $\boldsymbol{p}_2$) or ($\boldsymbol{p}_1^*$, $\boldsymbol{p}_2^*$) represented with reference to a polar coordinate frame. (3) Synergy variables are transformed into the A-A ratio $\Delta\boldsymbol{r}$ as follows:

$$\Delta\boldsymbol{r} = w_1\boldsymbol{q}_1 + w_2\boldsymbol{q}_2 \tag{2.54}$$

$$= w_1\boldsymbol{A}^+\boldsymbol{p}_1 + w_2\boldsymbol{A}^+\boldsymbol{p}_2 \tag{2.55}$$

Fig. 2.12 (continued) linearly superposing the weighted muscle synergies ($\Delta\boldsymbol{r} = \boldsymbol{r} - \bar{\boldsymbol{r}} = w_1\boldsymbol{q}_1 + w_2\boldsymbol{q}_2$), where $\bar{\boldsymbol{r}}$ is an average vector of the A-A ratio and $\boldsymbol{p}_1 = \boldsymbol{A}\boldsymbol{q}_1$, $\boldsymbol{p}_2 = \boldsymbol{A}\boldsymbol{q}_2$, and $\boldsymbol{p}_j$ (j=1,2) denote the synergy-force vector. The weights are uniquely determined by the synergy scores (w_1, w_2) which influence the direction of endpoint force. The deviation of the A-A ratio $\Delta\boldsymbol{r}$ may be transformed into the deviation of the endpoint force $\Delta\boldsymbol{F}$ by a linear mapping operator $\boldsymbol{A}$ ($\Delta\boldsymbol{F} = \boldsymbol{A}\Delta\boldsymbol{r}$). Two synergy-force vectors respectively point to the radial and angular directions at the endpoint in the polar coordinate system centered on the shoulder joint (see (**a**)). (**c**) The A-A sum $\boldsymbol{s}$ ($= [s_1, s_2, s_3, s_4]^T$) is another set of motor commands which influence endpoint stiffness. The combination of two central motor commands, the A-A ratio $\boldsymbol{r}$ and the A-A sum $\boldsymbol{s}$, generates peripheral commands for agonist-antagonist muscle pairs. (**d**) Motor command for each muscle. It is difficult to interpret its physical meaning without muscle synergy analysis on the basis of the A-A concept which corresponds to (2.52). The synergy-force vectors $\boldsymbol{p}_j$ ($j = 1, 2$) can be formally mapped onto the muscle synergies $\boldsymbol{q}_j$ with a linear transform matrix $\boldsymbol{A}^+$, where $\boldsymbol{A}^+$ is an inverse map, such as the pseudo-inverse matrix of $\boldsymbol{A}$. (4) Muscle commands are generated from the A-A ratio and A-A sum. Following (2.44) and (2.45), muscle activities can be derived from

$$m_{2j-1} = r_j s_j \tag{2.56}$$

$$m_{2j} = (1 - r_j) s_j \tag{2.57}$$

where j indicates the muscle-pair index. Though this paper omits the details of the A-A sum, task variables include endpoint (hand) stiffness, and it is transformed into the A-A sum, i.e., another variable to represent the activity of an antagonistic muscle pair. This framework, based on the analysis results of EMG signals, is expected to be effective for the control scheme of musculoskeletal robots, because of its simplification. This framework helps to handle the redundancy problem since it uses a small number of variables to generate behavior movements. Moreover, the framework clarifies the function of synergies in producing forces, an important aspect of motor control.

2.4 Conclusion

Analysis of muscle activities helps to investigate movement properties, and it may benefit musculoskeletal robot control. However, it is difficult to investigate muscle activities and measure each muscle force to implement muscle synergies. Moreover, the relationship between muscle activities and task-relevant variables is still unclear. Therefore, we have examined the EMGs (or the collection of input to each muscle) and hand force (or output of muscle-group force) and estimated the input-output model. In this chapter, we have characterized and evaluated the properties of synergies extracted from EMG signals in a force-producing task. First, we have assessed the muscle synergies extracted from the A-A ratios of measured EMGs. The analysis of muscle synergies based on A-A ratios is more interesting than that based on original EMGs. It helps to elucidate the physical meaning of muscle synergies more clearly. Two synergies are identified: one relating to the radial movement and the other resulting in the angular movement in a polar coordinate system centered on the shoulder joint. In addition, on the basis of the outcomes of muscle synergies regarding the A-A concept (a comparable form to the EPH), we have proposed a novel framework of human motor control that would be useful for explaining the mechanism of movement control. The basic idea of this framework is originated in the control of a redundant robot system with multiple PAMs. Using the analogy between the biological system and the robotic system with PAMs, we have challenged to a mystery of human motor control from the standpoint of *robotic-inspired* motor control. This framework suggests a promising solution to the redundancy problem in motor control, since it requires a small number of control variables to generate behavioral movements. Finally, it is noteworthy that this framework has a potential to extend its control method to more difficult tasks and more complex robot systems by modifying task variables according to specific desired tasks and by modifying PCs according to specific system specifications.

Acknowledgements The second half of this chapter is modified from the original (Pham et al. 2014), with kind permission from Taylor & Francis.

Exercises

1. Find a couple of examples of the Bernstein problem of degrees of freedom in everyday movement. The motor redundancy problem may exist not only in a joint space but also in a muscle space.
2. The EPH has been a controversial theory of motor control for about half a century. Survey this remarkable theory and summarize its pros and cons.
3. Verify Eq. (2.8) from the input-output relationship of energy in a single PAM.
4. Considering the two-joint arm model with three antagonistic pairs of PAMs in Fig. 2.6, derive Eq. (2.39) which describes the relationship among EPs, A-A ratios, and A-A sums.
5. Consider the role of one motor command, the A-A sum, and describe what kinds of tasks its control is effective for.
6. Motor synergies may be described in terms of three features: sharing, flexibility/stability (error compensation), and task dependence (Latash 2008b). PCA is a powerful technique for detecting the first feature of synergies. What methods are available for testing the second feature of synergies?

References

Ariga, Y., Pham, H., Uemura, M., Hirai, H., Miyazaki, F.: Novel equilibrium-point control of agonist-antagonist system with pneumatic artificial muscles. In: Proceedings of the IEEE International Conference on Robotics and Automation (ICRA2012), Saint Paul, pp. 1470–1475 (2012a)

Ariga, Y., Maeda, D., Pham, H., Uemura, M., Hirai, H., Miyazaki, F.: Novel equilibrium-point control of agonist-antagonist system with pneumatic artificial muscles: II. Application to EMG-based human-machine interface for an elbow-joint system. In: Proceedings of the IEEE International Conference on Intelligent Robots and Systems (IROS2012), Vilamoura-Algarve, pp. 4380–4385 (2012b)

Artemiadis, P.K., Kyriakopoulos, K.J.: EMG-based control of a robot arm using low-dimensional embeddings. IEEE Trans. Robot. **26**(2), 393–398 (2010)

Bernstein, N.: The Coordination and Regulation of Movements. Pergamon Press, Oxford (1967)

Bizzi, E., d'Avella, A., Saltiel, P., Tresch, M.: Modular organization of spinal motor systems. Neuroscientist **8**(5), 437–422 (2002)

Bizzi, E., Cheung, V.C.K., d'Avella, A., Saltiel, P., Tresch, M.: Combining modules for movement. Brain Res. Rev. **57**(1), 125–133 (2008)

Chou, C., Hannaford, B.: Measurement and modeling of Mckibben pneumatic artificial muscle. IEEE Trans. Robot. Autom. **12**(1), 90–102 (1996)

d'Avella, A., Bizzi, E.: Shared and specific muscle synergies in natural motor behaviours. Proc. Natl. Acad. Sci. USA **102**(8), 3076–3081 (2005)

d'Avella, A., Saltiel, P., Bizzi, E.: Combinations of muscle synergies in the construction of a natural motor behavior. Nat. Neurosci. **6**(3), 300–308 (2003)

Feldman, A.G.: Functional tuning of the nervous system with control of movement or maintenance of a steady posture, II: controllable parameters of the muscle. Biophysics **11**, 565–578 (1966)

Feldman, A.G., Levin, M.F.: The equilibrium-point hypothesis – past, present and future. In: Progress in Motor Control, A Multidisciplinary Perspective, pp. 699–726. Springer, Dordrecht (2008)

Feldman, A.G., Adamovich, S.V., Ostry, D.J., Flanagan, J.R.: The origin of electromyograms – explanations based on the equilibrium-point hypothesis. In: Multiple Muscle Systems, Biomechanics and Movement Organization, pp. 195–213. Springer–Verlag, New York Inc. (1990)

Fujimoto, S., Ono, T., Ohsaka, K., Zhao, Z.: Modeling of artificial muscle actuator and control design for antagonistic drive system. Trans. Jpn. Soc. Mech. Eng. **73**(730), 1777–1785 (2007) (in Japanese)

Giszter, S., Patil, V., Hart, C.: Primitives, premotor drives, and pattern generation: a combined computational and neuroethological perspective. Prog. Brain Res. **165**, 323–346 (2007)

Gottlieb, G.L.: Muscle activation patterns during two types of voluntary single-joint movement. J. Neurophysiol. **80**(4), 1860–1867 (1998)

Hislop, H., Montgomery, J.: Daniels Worthingham's Muscle Testing: Techniques of Manual Examination, 8th edn. Saunders, St. Louis (2007)

Ivanenko, Y.P., Popplele, R.E., Lacquaniti, F.: Five basic muscle activation patterns account for muscle activity during human locomotion. J. Physiol. **556**(1), 267–282 (2004)

Jolliffe, I.: Principal Components Analysis. Springer, New York (1986)

Kagawa, T., Fujita, T., Yamanaka, T.: Nonlinear model of artificial muscle. Trans. Soc. Inst. Control Eng. **29**(10), 1241–1243 (1993) (in Japanese)

Latash, M.L.: Evolution of motor control: from reflexes and motor programs to the equilibrium-point hypothesis. J. Hum. Kinet. **19**(19), 3–24 (2008a)

Latash, M.L.: Synergy. Oxford University Press, Oxford/New York (2008b)

Milner, T.E., Cloutier, C., Leger, A.B., Franklin, D.W.: Inability to activate muscles maximally during cocontraction and the effect on joint stiffness. Exp. Brain Res. **107**(2), 293–305 (1995)

Pham, H., Kimura, M., Hirai, H., Miyazaki, F.: Extraction and implementation of muscle synergies in hand-force control. In: Proceedings of the IEEE International Conference on Robotics and Automation (ICRA2011), Shanghai, pp. 3658–3663 (2011)

Pham, H., Ariga, Y., Tominaga, K., Oku, T., Nakayama, K., Uemura, M., Hirai, H., Miyazaki, F.: Extraction and implementation of muscle synergies in neuro-mechanical control of upper limb movement. Adv. Robot. **28**(11), 745–757 (2014)

Shin, D., Kim, J., Koike, Y.: A myokinetic arm model for estimating joint torque and stiffness from EMG signals during maintained posture. J. Neurophysiol. **101**(1), 387–401 (2009)

Ting, L.H.: Dimensional reduction in sensorimotor systems: a framework for understanding muscle coordination of posture. Prog. Brain Res. **165**, 299–321 (2007)

Tondu, B., Lopez, P.: Modeling and control of McKibben artificial muscle robot actuators. IEEE Control Syst. Mag. **20**(2), 15–38 (2000)

Chapter 3
Mechanism for Cognitive Development

Yukie Nagai

Abstract This chapter describes computational approaches to a new understanding of the mechanisms of cognitive development. There are two important factors for proper development: inherent mechanisms for infants to bootstrap their development and environmental scaffolding to facilitate development. After describing our key ideas, three case studies in cognitive developmental robotics are presented: a computational model for the emergence of the mirror neuron system, a developmental mechanism for joint attention, and an analysis of the dynamical information flow in infant-caregiver interaction. Lastly, the potential of our models to reveal the mechanisms of developmental disorders are discussed.

Keywords Cognitive developmental robotics (CDR) • Contingency/contingency learning • Self and others • Caregivers' scaffolding/Caregiver's scaffolding • Zone of proximal development (ZPD) • Mirror neuron system (MNS) • Joint attention • Social contingency • Information flow • Autism spectrum disorder (ASD) • Prediction error

3.1 Introduction

What makes us human? How do humans acquire cognitive abilities? These fundamental questions have been addressed by many researchers. Developmental psychologists have been investigating behavioral changes in human infants. For example, Jean Piaget, a Swiss developmental psychologist, conducted a pioneering study that built the foundation for the paradigms of current psychological experiments (Bremner 1994; Butterworth and Harris 1994; Piaget 1952). He closely observed his three children and analyzed their behaviors, resulting in a theory about four stages of development: (a) the sensorimotor stage through which infants learn to acquire sensorimotor mapping on the basis of their experiences, (b) the preoperational stage through which infants start manipulating symbols, and (c-d) the

Y. Nagai (✉)
Graduate School of Engineering, Osaka University, Suita, Osaka, Japan
e-mail: yukie@ams.eng.osaka-u.ac.jp

M. Kasaki et al. (eds.), *Cognitive Neuroscience Robotics A*,
DOI 10.1007/978-4-431-54595-8_3

concrete- and formal-operational stages in which infants develop logical thinking and abstract reasoning. While Piaget focused on the developmental change in children, Lev S. Vygotsky, a Russian psychologist, emphasized the role of environmental scaffolding (Berk and Winsler 1995; Vygotsky 1978). Children who cannot achieve a goal on their own often receive assistance from caregivers. Caregivers, for example, point to goals, embody children's actions (i.e., direct children's body so that they can experience perceptual information in the ways required for actions), and educate their attention (Zukow-Goldring and Arbib 2007). The environment children interact with is also shaped by caregivers. Such caregivers' support allows children to learn how to achieve their goals through guided experiences. Vygotsky's famous concept, called the "Zone of Proximal Development" (ZPD), describes the difference between what children can achieve *without* scaffolding and what they can do *with* scaffolding. He suggests that caregivers' scaffolding should be given within the ZPD, which is another important factor in infant development.

Neuroscientists have been investigating the development of infants' brain function. Recent advances in neuroimaging techniques enable researchers to examine the brain's activity in response to various stimuli. For example, Marshall et al. (2011) have shown that the magnitude of EEG (i.e., electroencephalography) desynchronization to action perception and production appears to be smaller for infants than for adults and older children. Shibata et al. (2012) have reported that the brains of newborns respond more broadly to tactile stimuli than to visual and auditory stimuli. These findings provide new insights into how differently infants and adults perceive the world and how infants gradually structure their brain mechanism through development.

In contrast to the two research fields mentioned above, cognitive developmental robotics (CDR) is a relatively new area investigating human development (Asada et al. 2001, 2009). CDR intends to fill the gap between psychology and neuroscience by means of a synthetic approach. Researchers in CDR design computational models that simulate the brain functions of infants and let robots equipped with these models interact with an environment. Their experiments, in which robots show a similar/dissimilar developmental process to/from infants, tell how and why infants do/do not acquire cognitive abilities. The advantages of this synthetic approach over behavioral and neural research include the embodiment of robots and their social interaction with the environment. The embodiment allows researchers to investigate the link between infants' brain activities and their behaviors, whereas they are studied separately in neuroscience and psychology. The embodiment further enables computational models to be grounded in social interaction with the environment. Studying the dynamical structure of social interaction and its change over development leads to a better understanding of the role of the ZPD.

This chapter first describes core issues for understanding the mechanisms of cognitive development. Two key ideas for cognitive development are introduced: contingency learning as an inherent mechanism used by infants and caregivers' scaffolding to facilitate development. Then, three case studies in CDR are presented to examine these key ideas: a computational model for the emergence of the mirror neuron system (MNS), a developmental mechanism for joint attention, and an analytical study of the dynamics of infant-caregiver interaction. These studies provide new insights into the mechanisms behind infant development. Lastly, a new hypothesis about developmental disorders, such as the autism spectrum disorder (ASD), is discussed. Our computational approach has the potential to reveal the mechanism of ASD.

3.2 Two Key Issues in Understanding Cognitive Development

There are two fundamental questions for understanding the mechanisms of cognitive development (Asada et al. 2001):

1. What inherent abilities should infants have for their development?
2. What environmental assistance is necessary for their proper development?

The first issue concerns the innate abilities of infants. To interact with their environment and acquire new skills, infants must be endowed with certain abilities to perceive, act, and memorize. Curiosity, for example, drives exploration of the environment. The ability to learn from experiences is required to obtain new skills. In other words, lack of or modification in such abilities may cause developmental disorders. Section 3.2.1 introduces the concept of contingency learning as a key inherent ability for infants.

The second issue concerns environmental scaffolding. To lead and accelerate proper development, infants need to receive certain assistance from the environment. We suggest that caregivers play an important role in facilitating infant development. As suggested by Vygotsky, caregivers' scaffolding given within the ZPD has the power to enable infants to experience what they cannot do on their own and thus facilitate their learning (Berk and Winsler 1995; Vygotsky 1978). Section 3.2.2 explains the mechanisms of caregivers' scaffolding and its effects on infant development.

3.2.1 Contingency Learning as Inherent Mechanism of Infants

What inherent abilities should infants have to bootstrap their cognitive development? What enables infants to acquire cognitive skills? We suggest that contingency learning is one of the core mechanisms for development.

From a computational point of view, contingency C_i of the i-th event is defined as a conditional probability between the perceptual state $S_i(t)$ of the event and the j-th action $A_j(t)$:

$$C_i = P\left(S_i(t+1) \mid S_i(t),\ A_j(t)\right), \tag{3.1}$$

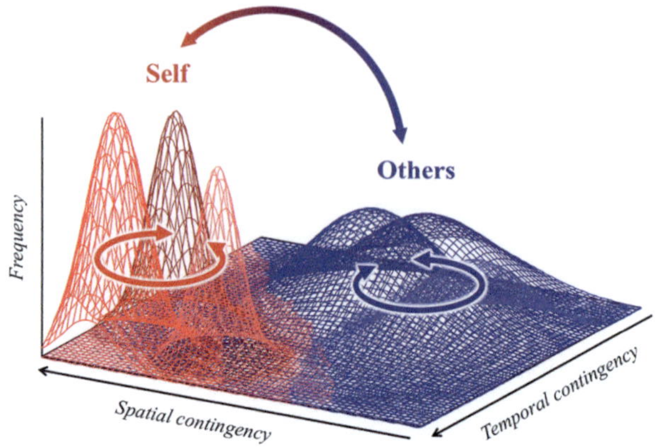

Fig. 3.1 The self and others defined by spatiotemporal contingency. Events concerning the self exhibit perfect contingency in terms of both space and time, whereas events related to others show lower contingency. Self/other recognition and social interaction are formulated as the coordination within and between the self and others

where t denotes a time. The higher the probability, the stronger the contingency of the i-th event. Figure 3.1 illustrates how the *self* and *others* are defined by contingency. The recognition of the self and others is considered as the basis for cognitive development. The $x - y$ plane represents the relationship between spatial and temporal contingency, whereas the z axis indicates the frequency of events. Spatiotemporal contingency means where and when events produce a state change. Clusters represented as Gaussian functions are mainly divided into two groups: clusters for the self in the left corner and clusters for others in the right corner. Events concerning the self exhibit nearly perfect contingency in terms of both space and time. For example, the visual state of our own body always produces the same state change in response to a certain motor command (e.g., opening the elbow joint always stretches the arm). In contrast, events concerning others exhibit a lower contingency and frequency in terms of both space and time. The responses of other individuals to an action produced by the self may vary depending on the context (e.g., waiving one's arm often, but not always, induces the greeting "goodbye" from other individuals).

Our key idea is that the goal of cognitive development is to identify the contingency map shown in Fig. 3.1 through experiences. In what follows, we explain the developmental change in contingency and the cognitive abilities that infants thereby acquire through development (see Fig. 3.2):

(a) To detect and control the self's body

Recognition of the self is a cornerstone for development. As mentioned above, the self's body has nearly perfect contingency and is thus distinguished from the environment:

$$C_{\text{self}} \approx 1.0 \tag{3.2}$$

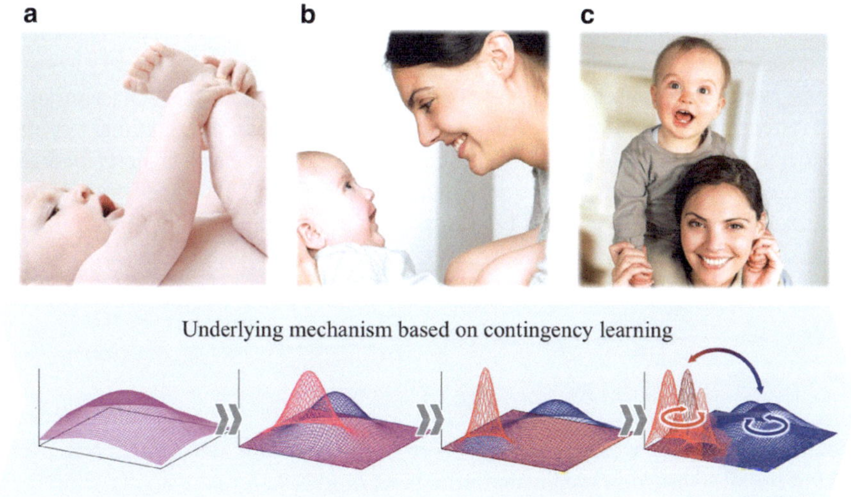

Fig. 3.2 Cognitive development of infants and underlying mechanism based on contingency. (**a**) The process of detecting and controlling the self identifies a cluster with perfect contingency. (**b**) Discriminating the self from others differentiates two clusters with different contingencies. (**c**) Communicating with others further segments clusters into behaviors and detects clusters of other individuals with higher contingencies

However, it is suggested that newborns cannot yet recognize the self. Their world is not differentiated, and thus they need to explore their own body by closely gazing at their hand (i.e., hand regard), putting their hand and foot into their mouth (i.e., double touch), and so on (Bremner 1994; Butterworth and Harris 1994). Experiencing their bodies through motion and perception, especially multimodal perception, enables them to learn to detect the self. A change from an undifferentiated cluster (the leftmost graph in Fig. 3.2) to two different clusters (the second and third graphs from the left in Fig. 3.2) illustrates this developmental process.

(b) To discriminate the self and others

Recognition of the self and recognition of others are two sides of the same coin. While learning to detect the self, infants also learn to detect other individuals, who have a lower but nonzero contingency:

$$C_{\text{others}} < C_{\text{self}} \quad \text{and} \quad C_{\text{others}} > 0.0 \tag{3.3}$$

The developmental process of self/other discrimination is represented as a separation of one cluster (i.e., non-differentiated self and others) into two clusters (i.e., differentiated self and others) in the contingency space. The change from the left to the center in Fig. 3.2 depicts this development as well as (a). As the cluster for the self (a taller and narrower one) becomes visible, the cluster for others (a shorter and wider one) comes to be distinguished from the self.

(c) To communicate with others

The development of social communication is the third step in contingency learning. Among various responses from other individuals, social behaviors such as imitation, joint attention, and language use, produce relatively higher contingency than nonsocial events. The development of social communication is described as a process of further classifying the self's and others' clusters into finer-grained ones, each of which corresponds to a type of behavior. The rightmost graph in Fig. 3.2 depicts this process:

$$C_{\text{others}} = \frac{1}{N+1} \sum C_{\text{others}_n} \quad \text{where} \quad \begin{cases} C_{\text{others}_0}, \ldots, C_{\text{others}_m} & \approx 0.9 \\ C_{\text{others}_{m+1}}, \ldots, C_{\text{others}_N} & \approx 0.0 \end{cases} \tag{3.4}$$

This definition indicates that the 0- to m-th events produced by other individuals are socially meaningful, whereas the $m + 1$- to N-th events are nonsocial. Thus, establishing social interaction means that infants learn to discriminate highly contingent behavior produced by other individuals from their noncontingent behavior and to respond only to the former. The process of refining the contingency clusters and coordinating them further continues to drive higher-level cognitive development.

3.2.2 Caregivers' Scaffolding to Facilitate Infant Development

What can facilitate contingency learning of infants? Can they develop without environmental support? We suggest that caregivers' scaffolding plays an important role in guiding infant development.

Caregivers often provide contingent reactions to infants so as to enable infants to learn about the environment (Csibra and Gergely 2009; Gergely et al. 2007). They, for example, smile at and talk to infants when infants achieve a goal or properly respond to caregivers. Such positive reward is supposed to motivate infants to learn their experiences and assist them in identifying the contingency map. It is also known that caregivers sometimes imitate infants (Catmur et al. 2009; Heyes 2010). Caregivers reproduce infants' utterance and copy infants' actions (e.g., facial expressions) to establish very first communication with them (especially when they are young). The experiences of being imitated enable infants to detect the correspondence between the self and others, which is an important step in contingency learning.

It has also been suggested that caregivers directly shape infants' experiences and learning. For example, caregivers educate infants' attention and embody their actions so as to enable infants to have experiences beyond their capabilities (Zukow-Goldring and Arbib 2007). According to Vygotsky, this scaffolding given within the ZPD encourages infants to learn new skills from their guided experiences. Other examples of caregivers' scaffolding are infant-directed speech (IDS) (Fernald and Simon 1984; Kuhl et al. 1997) and infant-directed action (IDA) (Brand et al.

2002; Rohlfing et al. 2006). IDS and IDA are characterized by exaggeration and simplification of caregivers' behaviors. For example, caregivers stretch the contour of their speech (Fernald and Simon 1984), expand the vowel space (Kuhl et al. 1997) and the trajectory of an action (Brand et al. 2002; Rohlfing et al. 2006), and so on, which are assumed to facilitate infant learning. Although there is a wide variety of caregivers' scaffolding, the important characteristic of it is that the degree of scaffolding is adjusted to fall within the ZPD.

The following sections describe the author's work to investigate the roles of contingency learning and caregivers' scaffolding in cognitive development. Computational models incorporating the above ideas are presented.

3.3 Emergence of MNS Through Self/Other Discrimination

3.3.1 *MNS and Its Development*

Mirror neurons and the MNS discharge both when a person is executing an action and when he/she is observing the same action performed by other individuals (Gallese et al. 1996; Rizzolatti and Sinigaglia 2008; Rizzolatti et al. 2001). Many researchers in neuroscience, developmental psychology, and even robotics have investigated various functions of the MNS. For example, the MNS enables us to understand the intention of other people's actions, to imitate the actions, and so on (e.g., Umiltà et al. 2001), which lead to the development of higher cognitive functions. However, the origin of the MNS is still an open question. How it develops before/after birth has not been fully understood because of the limitations in neuroimaging techniques.

We support the hypothesis that the MNS develops through sensorimotor learning in infancy. Heyes and her colleagues (Catmur et al. 2009; Heyes 2010) propose the model called "associative sequence learning," according to which infants acquire the MNS as a by-product of sensorimotor development. Their model supposes that infants are often imitated by caregivers. The experience of being imitated enables infants to map the caregivers' action onto the infants' motor command, which produces the MNS. Their model, however, does not explain how infants can distinguish the self from others. Discrimination between the self and others is as important as detection of the correspondence between them for social interaction (e.g., turn taking).

3.3.2 *Computational Model for Emergence of MNS*

We have previously proposed a computational model for the emergence of the MNS (Kawai et al. 2012; Nagai et al. 2011). A key idea of our model is that perceptual

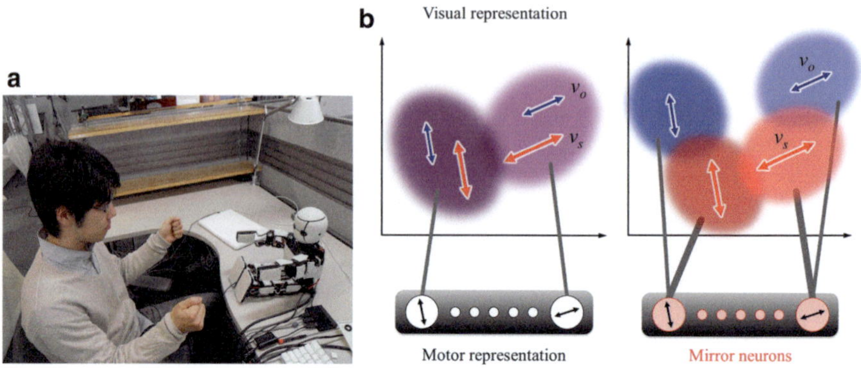

Fig. 3.3 Computational model for the emergence of the MNS (Modified from Nagai et al. 2011). (**a**) An infant-like robot interacting with a human caregiver. The robot learns sensorimotor coordination through body babbling while receiving contingent responses from the caregiver. (**b**) A learning model for sensorimotor coordination: the early stage (*left*) and the later stage (*right*) of development. The MNS emerges through sensorimotor learning synchronized with visual development

development leads to a gradual discrimination between the self and others while maintaining the correspondence between them.

Figure 3.3a, b show an experimental setting and our proposed model, respectively. An infant-like robot learns sensorimotor mapping through body babbling while interacting with a human caregiver. The two figures in Fig. 3.3b illustrate the association between motor representation (the lower layer) and visual representation (the upper layer) of the robot. In the early stage of learning (left side of Fig. 3.3b), the robot cannot discriminate between its own motion v_s and others' motion v_o because of a lower acuity of visual perception. The difference between the self's contingency and others' contingency, which was discussed in Sect. 3.2.1, is considerably too small to be separated into two clusters. Therefore, the visual clusters containing both the self's and others' motions are associated with the same motor neurons by Hebbian learning. Note that this stage corresponds to the leftmost part of Fig. 3.2. In the later stage of learning (right side of Fig. 3.3b), the robot improves the acuity of its visual perception and thus discriminates the self from others. The clusters previously contained both the self's and others' motions but now separate into two clusters owing to their inherent difference in contingency. The important point here is that the self's clusters and the corresponding others' clusters are associated with the same motor neurons. The previously acquired association between the motor neurons and the non-differentiated visual clusters has been maintained through development. Thus, the motor neurons activate not only when the robot produces an action but also when it observes the same action performed by the caregiver. This result exhibits the function of the MNS.

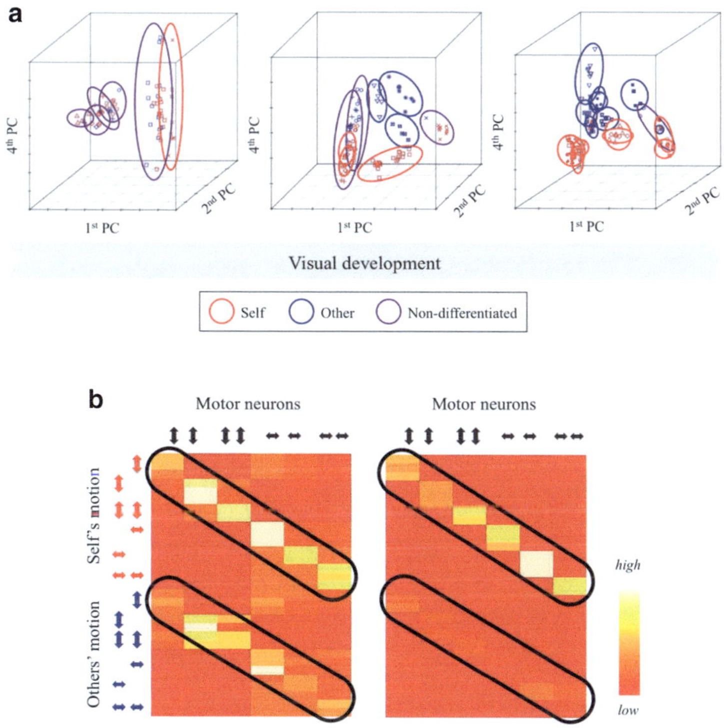

Fig. 3.4 Development of visual representation and the MNS emerging through development (Modified from Nagai et al. 2011). (**a**) Self/other discrimination through visual development. As the acuity of visual perception improves, the clusters in the visual space separate into the self and others. (**b**) The MNS acquired in sensorimotor coordination. The *left* shows a result *with* visual development, and the *right* shows a result *without* visual development. The stronger connection between others' motions and the motor neurons in the *lower left* indicates the function of the MNS

3.3.3 Roles of Contingency Learning and Caregiver's Scaffolding in Emergence of MNS

Figure 3.4 shows the results of our experiments on (a) self/other discrimination in the visual space and (b) the acquisition of connecting weights between visual and motor representations. During learning, the robot randomly produced one of six types of motions, while the caregiver imitated them at a level of 30 % with some delay and variation.

In the early stage of development, most of the visual clusters contain both the self's and others' motions, as shown in the leftmost part of Fig. 3.4a. At this stage, the robot cannot differentiate between the self and others because of the low acuity of its visual perception. As the acuity improves, the robot starts differentiating between them. Some of the clusters in the middle of Fig. 3.4a contain either the self's or others' motions. Finally, the self and others are clearly separated in the rightmost of Fig. 3.4a. The clusters for the self are found in the lower part, whereas the clusters for others are in the upper portion. The difference in contingency between the self and others enables the robot to discriminate between them.

The connecting weights between visual representation and motor representation verify the acquisition of the MNS. Figure 3.4b shows the acquired connecting weights under two conditions: *with* (left) and *without* (right) visual development. Without visual development, the robot used the highest acuity of visual information over learning. The rows indicate the clusters of visual representation, and the columns represent the six motor neurons. For example, the leftmost column shows the connecting weight for the motor neuron controlling up-and-down right-hand movement (indicated by a small arrow). The lighter the color, the stronger the connection. Both results verify that a strong contingency toward the self's motion has successfully been acquired regardless of visual development. The stronger weights from the top-left corner to the middle-right show a proper association regarding the self's motion. It is natural that the robot acquires this association because its motions have perfect contingency. In contrast, the association between the motor neurons and others' motions, which is related to the MNS, reveals a significant difference between the two conditions. Only when the robot was equipped with visual development (left side of Fig. 3.4b), did it acquire the MNS. The stronger diagonal connection in the lower part represents the MNS. These results suggest that contingency learning synchronized with perceptual development can lead to the emergence of the MNS.

Our results emphasize the importance of caregivers' scaffolding as well as contingency learning. In our experiments, the caregiver imitated the robot's motion at 30 %. Our preliminary experiment showed that less or no imitation by the caregiver makes it difficult to acquire the MNS. Implementing only the visual development without the caregiver's scaffolding is not sufficiently powerful to enable the robot to detect the self-other correspondence. This finding empirically supports our key idea that both inherent mechanisms for robots and caregivers' scaffolding should be properly designed.

3.4 Development of Joint Attention Based on Social Contingency

3.4.1 Development of Joint Attention in Infants

Joint attention is the process of looking at an object that someone else is looking at by following the direction of his/her gaze (Moore and Dunham 1995; Scaife and Bruner 1975). The ability to achieve joint attention is believed to be a milestone in cognitive development, because it leads to the acquisition of various social abilities, such as language (Brooks and Meltzoff 2005) and theory of mind (Baron-Cohen 1995). Butterworth and Jarrett (1991) examined how the accuracy of gaze following improves during infancy. They suggested three stages of development: (a) the ecological stage in which infants look at a salient object while ignoring the exact direction of caregivers' gaze, (b) the geometric stage in which infants properly follow caregivers' gaze only when the target object is in the infants' view, and (c) the representational stage in which infants achieve joint attention even if the target object is outside their view for some time at the beginning.

Our interest is in how the mechanism of contingency learning bootstraps these stages of development and how caregivers contribute to the facilitation of learning. The following sections present our computational studies to address these questions.

3.4.2 *Contingency Learning for Joint Attention Development*

We suggest that contingency learning integrated with saliency-based attention enables robots to reproduce infant-like stages of development (Nagai et al. 2003). Saliency-based attention allows robots to explore the environment, whereas contingency learning detects a higher correlation in sensorimotor coordination when the robots achieve joint attention. As described in Sect. 3.2.1, social behaviors, such as joint attention, have higher contingency than nonsocial behaviors. Moreover, if a learning module has a limited capacity in its memory, it is expected that only higher correlations obtained through a success of joint attention are acquired in sensorimotor mapping, and rather than that all experiences of sensorimotor coordination (i.e., including both success and failure of joint attention) are memorized. The ability to express joint attention is therefore acquired as a stronger correlation in sensorimotor coordination without receiving any social feedback.

Figure 3.5a, b shows our experimental setup and proposed model, respectively. The robot first obtains a camera image $\boldsymbol{I}$ and then outputs a motor command $\boldsymbol{\Delta\theta}$ to rotate the camera head. The model consists of three modules:

- Visual attention makes the robot gaze at a salient object in terms of primitive features (e.g., color, edge, and motion),
- Learning with self-evaluation enables the robot to learn the sensorimotor mapping between a facial image of the caregiver and the gaze shift using a neural network, and
- A gate determines which output from the two modules should be selected to turn the robot's head.

An important point is that the robot does not receive any feedback from the caregiver. Instead, social contingency between the robot's bottom-up attention and the caregiver's attention directed mostly toward a salient object enables the robot to acquire a stronger sensorimotor correlation for joint attention. The gate plays another important role in reproducing the three stages of development. It gradually shifts the selection of the robot's output from bottom-up to contingency-based attention, resulting in the developmental shift from (a) to (c) in infants.

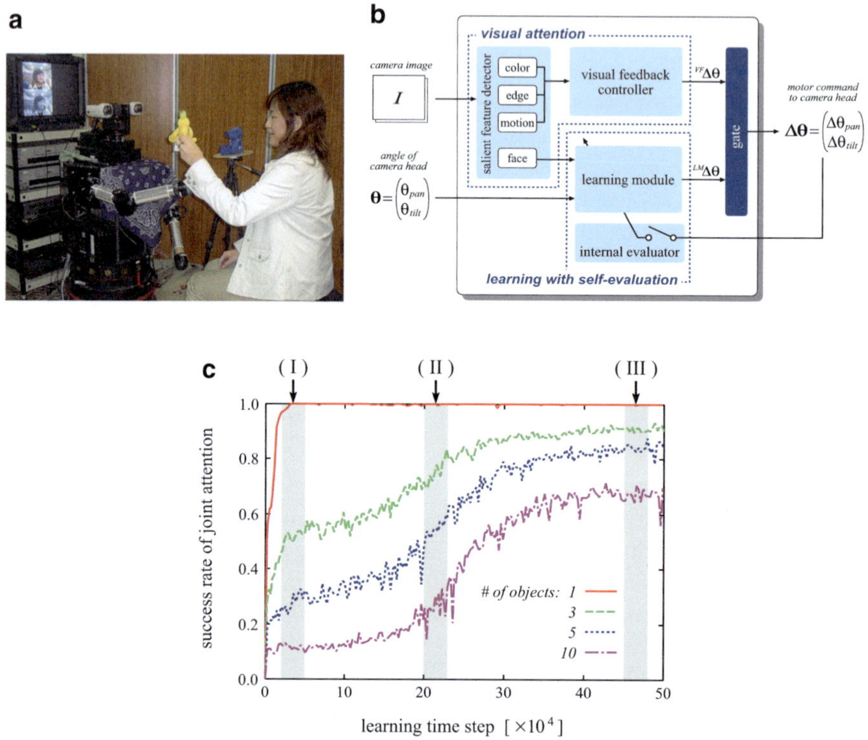

Fig. 3.5 Development of joint attention based on contingency learning (Adapted from Nagai et al. 2003). (**a**) A robot learning to achieve joint attention with a human caregiver. (**b**) A learning model for joint attention based on contingency learning combined with saliency-based attention. (**c**) Learning performance with different numbers of objects. The robot exhibits the three stages of development like infants at phases I, II, and III

The results of the learning experiment are depicted in Fig. 3.5c. The curves plot the success rate of joint attention with different numbers of objects. Our robot was able to achieve high joint attention performance even when the number of objects increased to 10. We also found that the robot exhibited the three stages of development as they are observed in infants. In phase I, the robot gazed at a salient object on the basis of bottom-up attention and then started following the direction of the caregiver's gaze in phase II by utilizing the contingent relationship acquired in sensorimotor mapping. These two phases correspond to the first two stages of infant development. In addition, the robot achieved joint attention even when the object was placed outside the robot's view at the beginning of the interaction (in phase III); this behavior corresponds to the third stage of infant development. There results suggest that contingency learning combined with a mechanism of environmental exploration can autonomously develop the ability of social communication.

3.4.3 Caregiver's Scaffolding to Facilitate Joint Attention Development

Our second model focuses on the role of scaffolding in joint attention development. What type of scaffolding should be given to the robot in order to facilitate its learning? How can caregivers assist the development of joint attention? Inspired by the concept of the ZPD (Berk and Winsler 1995; Vygotsky 1978), we designed a computational model including a caregiver, who provides social feedback to a robot according to its learning progress (Nagai et al. 2006).

Figure 3.6a illustrates the basic idea of our model in the early stage (left) and the later stage (right) of development. Similar to the previous model, this model enables the robot to obtain a facial image of the caregiver and then generate a gaze shift as motor output using a neural network. An important factor of this model is that the

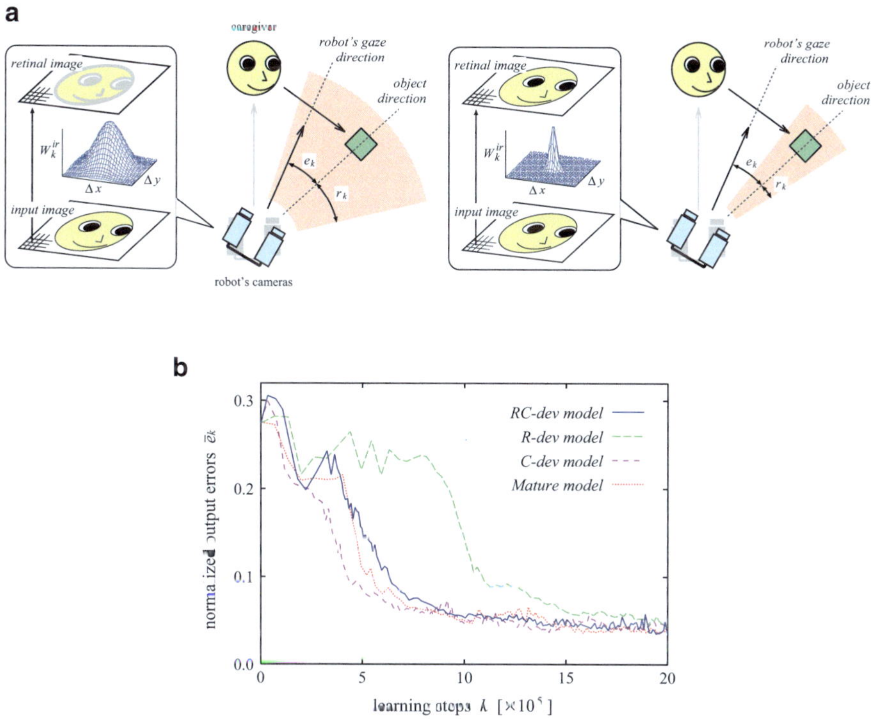

Fig. 3.6 Development of joint attention facilitated by caregivers' scaffolding (Adapted from Nagai et al. 2006). (**a**) Basic concept of a learning model with caregivers' scaffolding: the early (*left*) and later (*right*) stages of development. The *fan-shaped* area surrounding the target object indicates the threshold r_k for a reward to be given to the robot. r_k decreases as the output error of the robot e_k decreases to facilitate learning. (**b**) Learning performance under four conditions: RC-dev and C-dev are models incorporating caregiver's scaffolding, whereas R-dev and matured models have no scaffolding. Learning is accelerated when the caregiver provides proper scaffolding

caregiver provides a reward to the robot; the criterion for a positive/a negative reward changes as learning advances. The fan-shaped area surrounding the target object in Fig. 3.6a represents the threshold r_k for a reward at the k-th learning step, whereas e_k represents the error between the direction of the robot's gaze and that of the object. If the output of the robot falls within the threshold (i.e., $e_k \leq r_k$), the caregiver gives the robot a positive reward. Otherwise (i.e., $e_k > r_k$), the caregiver gives no reward. The reward is then used for updating the neural network. The connecting weights of the network are maintained when a reward is given, whereas the weights are slightly modified with random values when no reward is given. Here, the threshold r_k is defined using the average of the output error of the robot $\bar{e}_{k-1}$ in the last several trials:

$$r_k = \bar{e}_{k-1} - \epsilon \quad (\epsilon\text{: a small value}), \tag{3.5}$$

so that the robot learns a slightly advanced ability within the ZPD.

Figure 3.6b shows the results of the learning experiment. We conducted the experiment under four different conditions. Here, only the results that focus on the effect of the caregiver's scaffolding are discussed. The RC-dev and C-dev models include the caregiver's scaffolding ('C' denotes caregiver), whereas the R-dev and matured do not. The threshold r_k was set to a small constant value over learning in the latter conditions. Comparing the former models (RC-dev and C-dev) with the latter models (R-dev and Matured) reveals the effect of the scaffolding on the robot's learning. The learning speed for the RC-dev is faster than that for the R-dev, and the same acceleration can be observed for the C-dev compared to the Matured. This indicates that the caregiver's adaptive reward facilitates learning. Other designs of the caregiver's adaptation, such as the use of a linearly decreasing threshold, appeared not to be as effective as Eq. (3.5). This suggests that the caregiver's scaffolding should be adjusted according to the advance in the robot's learning in order to maximize the effect of scaffolding.

3.5 Developmental Dynamics of Social Contingency in Infant-Caregiver Interaction

3.5.1 Dynamical Structure of Interaction

Interaction between an infant and a caregiver is a dynamic and bidirectional process (Cohn and Tronick 1988; Kaye and Fogel 1980). They send various signals to each other, which elicit and shape each other's reactions. For example, a caregiver's gaze indicates an interesting event in the environment and therefore motivates an infant to look at and share the event with the caregiver (i.e., joint attention). Hand movements also convey information to a partner. Both communicative gestures (e.g., pointing) and task-related movements (e.g., playing with a toy) influence a partner's reactions. IDA (Brand et al. 2002; Nagai and Rohlfing 2009; Rohlfing et al. 2006) and IDS (Fernald and Simon 1984; Kuhl et al. 1997), discussed in Sect. 3.2.2, are typical phenomena observed in caregivers' behavior. They can attract and guide infants' attention to important information (Koterba and Iverson 2009). An open question here is how the dynamics of infant-caregiver interaction change with infants' age. Uncovering the link between infants' contingency learning and caregivers' scaffolding in their dynamical interaction is important for designing continuous development.

3.5.2 Analysis of Information Flow in Infant-Caregiver Interaction

We measured the dynamics of infant-caregiver interaction by applying the information theoretic measure called "transfer entropy" (Nagai et al. 2012). Let U and V be variables that are approximated by stationary Markov processes of orders k and l, respectively. Transfer entropy (Schreiber 2000) defines the degree of influence of V on U by

$$T_{V \to U} = \sum p(u_{t+1}, u_t^{(k)}, v_t^{(l)}) \log \frac{p(u_{t+1}|u_t^{(k)}, v_t^{(l)})}{p(u_{t+1}|u_t^{(k)})}, \tag{3.6}$$

where u_{t+1} is the value of U at time $t+1$; $u_t^{(k)}$ and $v_t^{(l)}$ are $(u_t, \dots u_{t-k+1})$ and $(v_t, \dots v_{t-l+1})$, respectively, and $p(u_{t+1}|u_t^{(k)}, v_t^{(l)})$ is the transition probability between them. Usually, $k = l = 1$ is used for computational reasons.

Using the transfer entropy, we investigated which of the signals given by an infant or a caregiver influence the other person and how the degree of influence changes with infants' age. Figure 3.7a shows a scene captured from our experiment, where a caregiver demonstrates a cup-nesting task to an infant. The 3D body movement of the participants was measured by Kinect sensors. In addition, the target of the participants' gaze was coded by hand. Figure 3.7b illustrates four types of information flow (denoted by arrows) measured in infant-caregiver interaction. Each column represents the time series of motion data of the participants. Let $m_{i,t}$ and $M_{j,t}$ be the motion of an infant's i-th body part and a caregiver's j-th body part at time t, respectively. The four types of information flow are defined as follows:

(i) Influence of $M_{j,t}$ on $m_{i,t+1}$ (i.e., $T_{M_j \to m_i}$),
(ii) Influence of $m_{i,t}$ on $M_{j,t+1}$ (i.e., $T_{m_i \to M_j}$),
(iii) Influence of $m_{k,t}$ on $m_{i,t+1}$ ($i \neq k$) (i.e., $T_{m_k \to m_i}$), and
(iv) Influence of $M_{k,t}$ on $M_{j,t+1}$ ($j \neq k$) (i.e., $T_{M_k \to M_j}$).

The first two values measure *social contingency* between the participants: $T_{M_j \to m_i}$ represents how much an infant responds to a caregiver's motion and vice versa for $T_{m_i \to M_j}$. Higher values of $T_{M_j \to m_i}$ and/or $T_{m_i \to M_j}$ indicate more contingency between the infant and the caregiver. The latter two values represent *body coordination*

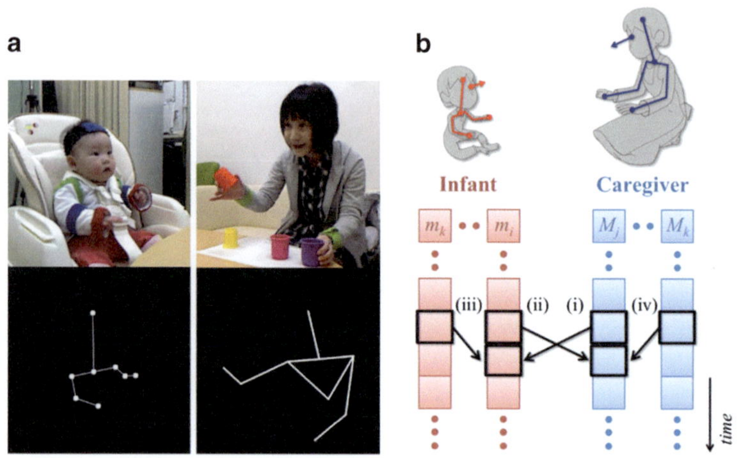

Fig. 3.7 Analysis of the dynamical structure of infant-caregiver interaction (Adapted from Nagai et al. 2012). (**a**) Interaction between an infant and a caregiver, whose body movement is measured by Kinect sensors. The caregiver demonstrates a cup-nesting task to the infant, who is watching and responding to the demonstration. (**b**) Four types of information flow in infant-caregiver interaction: (i–ii) and (iii–iv) denote social contingency and body coordination between and within participants, respectively

within each participant: $T_{m_k \to m_i}$ indicates the degree to which an infant's body part is coordinated with his/her other body parts, whereas $T_{M_k \to M_j}$ measures the coordination between body parts of a caregiver. Higher values of $T_{m_k \to m_i}$ and/or $T_{M_k \to M_j}$ imply a better ability of the infant and/or the caregiver to coordinate their body (e.g., hand-eye coordination and bimanual manipulation).

3.5.3 Development of Infants' Social Contingency Facilitated by Caregivers

Figure 3.8 shows the results of our analysis. Only two information flows ((a) $T_{M_j \to m_i}$ and (b) $T_{m_i \to M_j}$) are presented here. Lighter (pink or light blue) and darker (red or blue) bars show the results for younger (6–8 months old) and older (11–13 months old) infants, respectively. The labels under the bars denote the motion data to be focused on: "RHAND→gaze", for example, indicates the transfer entropy $T_{M_{\text{rhand}} \to m_{\text{gaze}}}$ from a caregiver's right hand to an infant's gaze.

Our first finding concerns the development of infants' social contingency. The significant difference in Fig. 3.8a suggests that older infants rely more strongly on caregivers' gaze than younger infants do, when they determine where to focus their attention (GAZE→gaze: $t(24) = 3.12, p < 0.01$). It is known that infants acquire the ability for gaze following and joint attention between 6 and 18 months of age (Butterworth and Jarrett 1991). Their gaze shift gradually becomes more contingent

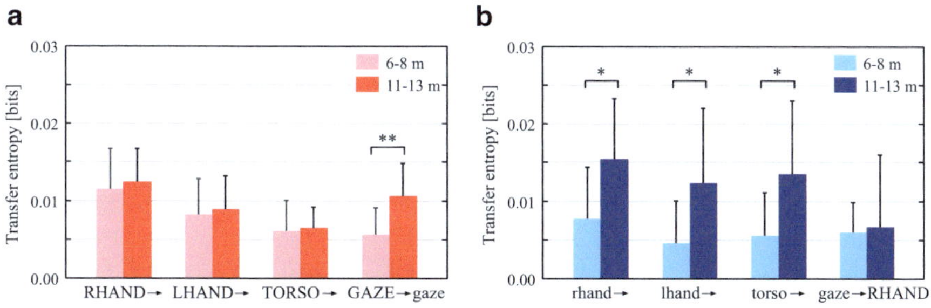

Fig. 3.8 Transfer entropy calculated in infant-caregiver interaction. Both (**a**) and (**b**) represent social contingency between the participants, which are denoted by (i) and (ii) in Fig. 3.7b, respectively. The *lighter* and *darker bars* are the results for younger and older infants, respectively (Adapted from Nagai et al. 2012)

on caregivers' gaze. Moreover, it has been suggested that the visual attention of younger infants relies more on bottom-up salience (e.g., motion and color) than on social signals (e.g., face and gaze) (Frank et al. 2009; Golinkoff and Hirsh-Pasek 2006). The higher transfer entropy for RHAND→gaze in younger infants indicates their preference for bottom-up salience because the caregivers' right hands move a lot during task demonstration.

Our second finding concerns caregivers' scaffolding. The significant differences in Fig. 3.8b indicate that caregivers increase social contingency in response to increases in the infants' age (rhand→RHAND: $t(24) = 2.53$, $p < 0.05$; lhand→RHAND: $t(24) = 2.47$, $p < 0.05$; torso→RHAND: $t(24) = 2.56$, $p < 0.05$). We suggest that this adaptation is caused by infant development. As shown in Fig. 3.8a, infants develop their social contingency from 6 to 13 months of age. This development enables caregivers to understand the intention and desire of infants' actions and thus to provide more social feedback to infants. We also found that the body coordination of infants and caregivers increases as infants grow. Taken together, our analysis reveals mutual development between infants and caregivers.

3.6 Conclusion and Discussion: Toward Understanding Mechanism of ASD

This chapter has described computational approaches to a new understanding of cognitive development. Our models and analyses have demonstrated the following:

- The ability of contingency learning bootstraps cognitive development in infants. Development of self/other recognition and social interaction, such as joint attention, can be achieved by detecting contingent relationships in sensorimotor coordination.

- Caregivers' scaffolding leads proper development of infants. Contingent reactions and social feedback provided by caregivers assist infants in detecting self/other correspondence (i.e., the MNS) and accelerating learning.

This section extends our idea to explain the mechanisms of ASD. ASD is a developmental disorder characterized by social deficits and communication difficulties (Baron-Cohen 1995). People with ASD also exhibit repetitive behaviors and interests. While traditional studies on ASD have focused on their disabilities in social aspects, recent studies emphasize their different way of processing sensorimotor information from the way adopted by typically developing people. Frith and Happé (1994) and Happé and Frith (2006) proposed the hypothesis called "weak central coherence." It is known that sensorimotor information is processed hierarchically in the human brain, where primitive features are extracted at a lower level and then gradually integrated to become more abstract at a higher level of cognition. The theory of weak central coherence suggests that a weaker ability to integrate information and/or a hyper ability to process primitive features cause social deficits in ASD. It is supposed that social interaction requires an access to the mental state of a partner, which is represented in a higher level of cognition. Therefore, a relative weakness in a higher level of cognition causes a deficit in social interaction.

Kumagaya, Ayaya, and their colleagues (Ayaya and Kumagaya 2008; Ayaya et al. 2013) conduct Tohjisha-kenkyu on ASD, that is, examine ASD from a phenomenological viewpoint. People suffering from ASD observe their own perceptual and motor experiences and try to interpret underlying mechanisms. The questions addressed in this study include the following: how the way they perceive the world is different from the way typically developing people do, what difficulties they have in recognizing their perception and determining their actions, what may cause such difficulties, and how they cope with the difficulties. The observational study leads to the hypothesis that the essential characteristic of ASD is the difficulty in integrating sensorimotor information rather than in social interaction, which is analogous to weak central coherence. Moreover, this study emphasizes the role of a prediction error as a cause of ASD. When recognizing perceptual information, people compute a prediction error between what they have expected to perceive and what they actually perceive (Blakemore et al. 1999). If the error is small, the perceptual information can be understood as a known event. If the error is large, the information cannot be understood and therefore is regarded as a new event. Kumagaya, Ayaya, and their colleagues (Ayaya and Kumagaya 2008; Ayaya et al. 2013) suggest that people with ASD may have a smaller or larger tolerance for prediction errors than typically developing people, which causes different recognition about the environment and thus deficit in social communication.

The author has been trying to formulate the hypothesis proposed by Kumagaya, Ayaya, and their colleagues, from a computational point of view (Nagai and Asada 2015). Figure 3.9 depicts the basic idea for the formulation. Assume that sensorimotor information perceived from the environment is represented as data points with respect to time and space (the upper graph of Fig. 3.9) and that linear regression is applied so as to recognize the perception by generating internal models

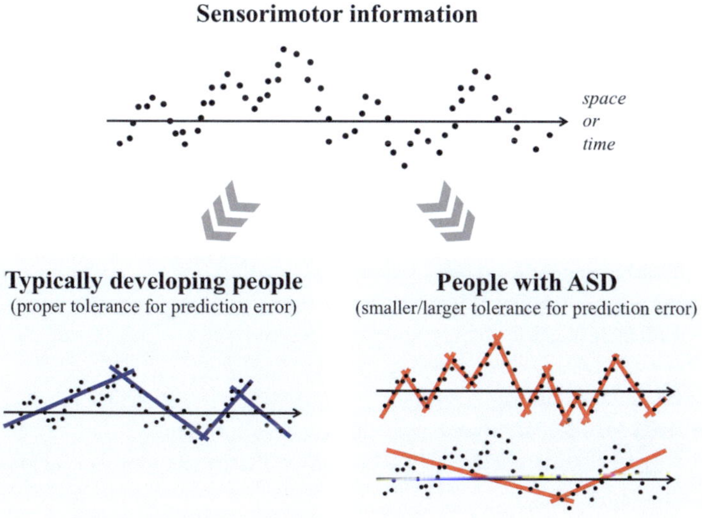

Fig. 3.9 How differently people with and without ASD recognize sensorimotor information. Typically developing people (*left*) are supposed to have a moderate tolerance for a prediction error and thus create an internal model with adaptability. People with ASD (*right*), on the other hand, show a smaller or a larger tolerance for a prediction error and thus generate a strict or a rough model without adaptability or reactivity. Such difference in their internal models results in a difficulty in social communication

of the world. According to Kumagaya, Ayaya, and their colleagues (Ayaya and Kumagaya 2008; Ayaya et al. 2013), people with ASD have a different tolerance (i.e., smaller or larger) from typically developing people. If people *without* ASD adopt a proper tolerance for prediction errors, they obtain the internal model as drawn by multiple lines in the lower left part of Fig. 3.9. Note that the lines do not exactly match the data points, indicating that people without ASD have adaptability to an environmental change. In contrast, people *with* ASD obtain different internal models as shown in the lower right of Fig. 3.9 because of their smaller or larger tolerance for prediction errors. A smaller tolerance generates a strict model with lower adaptability, whereas a larger tolerance generates a rough model with less reactivity; these two types may have analogy to hyperesthesia and hypoesthesia, respectively. This formulation makes it easier for us to understand why people with ASD have the difficulty in communicating with others. Because of their internal models that are different from typically developing people's, they cannot easily understand the intention and belief of other people (i.e., a deficit in theory of mind (Baron-Cohen 1995)). We aim to further investigate the mechanisms of ASD as well as those of typically developing people in order to better understand the principle of cognitive development.

Exercises

Answer the following questions:

1. Inherent abilities of infants
   What inherent abilities (other than contingency learning) do infants have for their cognitive development? What roles do these abilities play in their development?
2. Embodiment of infants
   This chapter has mainly discussed the development of cognitive functions. However, the bodies of infants grow as they develop. How does the growth of their bodies influence their cognitive development? What is the advantage of starting with a small size of bodies in infancy, if any?
3. Caregivers' scaffolding
   Are there types of caregivers' scaffolding which facilitate infant development in other ways than this chapter describes? How do those types of scaffolding assist infant development?

References

Asada, M., MacDorman, K.F., Ishiguro, H., Kuniyoshi, Y.: Cognitive developmental robotics as a new paradigm for the design of humanoid robots. Robot. Auton. Syst. **37**(2–3), 185–193 (2001)

Asada, M., Hosoda, K., Kuniyoshi, Y., Ishiguro, H., Inui, T., Yoshikawa, Y., Ogino, M., Yoshida, C.: Cognitive developmental robotics: a survey. IEEE Trans. Auton. Ment. Dev. **1**(1), 12–34 (2009)

Ayaya, S., Kumagaya, S.: Hattatsu Shougai Tojisha Kenkyu (in Japanese). Igaku-shoin, Tokyo (2008)

Ayaya, S., Kawano, T., Mukaiyachi, I., Tojisha-Kenkyukai, N., Ishihara, K., Ikeda, T., Kumagaya, S.: Tojisha Kenkyu no Kenkyu (in Japanese). Igaku-shoin, Tokyo (2013)

Baron-Cohen, S.: Mindblindness. MIT, Cambridge (1995)

Berk, L.E., Winsler, A.: Scaffolding Children's Learning: Vygotsky and Early Childhood Education. National Association for the Education of Young Children, Washington, DC (1995)

Blakemore, S.-J., Frith, C.D., Wolpert, D.M.: Spatio-temporal prediction modulates the perception of self-produced stimuli. J. Cogn. Neurosci. **11**(5), 551–559 (1999)

Brand, R.J., Baldwin, D.A., Ashburn, L.A.: Evidence for 'motionese': modifications in mothers' infant-directed action. Dev. Sci. **5**(1), 72–83 (2002)

Bremner, J.G.: Infancy. Wiley-Blackwell, Hoboken (1994)

Brooks, R., Meltzoff, A.N.: The development of gaze following and its relation to language. Dev. Sci. **8**(6), 535–543 (2005)

Butterworth, G., Harris, M.: Principles of Developmental Psychology. Lawrence Erlbaum Associates, Mahwah (1994)

Butterworth, G., Jarrett, N.: What minds have in common is space: spatial mechanisms serving joint visual attention in infancy. Br. J. Dev. Psychol. **9**, 55–72 (1991)

Catmur, C., Walsh, V., Heyes, C.: Associative sequence learning: the role of experience in the development of imitation and the mirror system. Philos. Trans. R. Soc. B Biol. Sci. **364**(1528), 2369–2380 (2009)

Cohn, J.E., Tronick, E.Z.: Mother-infant face-to-face interaction: influence is bidirectional and unrelated to periodic cycles in either Partner's behavior. Dev. Psychol. **24**(3), 386–392 (1988)

Csibra, G., Gergely, G.: Natural pedagogy. Trends Cogn. Sci. **13**(4), 148–153 (2009)

Fernald, A., Simon, T.: Expanded intonation contours in mothers' speech to newborns. Dev. Psychol. **20**(1), 104–113 (1984)
Frank, M.C., Vul, E., Johnson, S.P.: Development of infants' attention to faces during the first year. Cognition **110**(2), 160–170 (2009)
Frith, U., Happé, F.: Autism: beyond "theory of mind". Cognition **50**, 115–132 (1994)
Gallese, V., Fadiga, L., Fogassi, L., Rizzolatti, G.: Action recognition in the premotor cortex. Brain **119**, 593–609 (1996)
Gergely, G., Egyed, K., Király, I.: On pedagogy. Dev. Sci. **10**(1), 139–146 (2007)
Golinkoff, R.M., Hirsh-Pasek, K.: BabyWordsmith: from associationist to social sophisticate. Curr. Dir. Psychol. Sci. **15**(1), 30–33 (2006)
Happé, F., Frith, U.: The weak coherence account: detail-focused cognitive style in Autism spectrum disorders. J. Autism Dev. Disord. **36**(1), 5–25 (2006)
Heyes, C.: Where do mirror neurons come from? Neurosci. Biobehav. Rev. **34**(4), 575–583 (2010)
Kawai, Y., Nagai, Y., Asada, M.: Perceptual development triggered by its self-organization in cognitive learning. In: Proceedings of the 2012 IEEE/RSJ International Conference on Intelligent Robots and Systems, Vilamoura, pp. 5159–5164 (2012)
Kaye, K., Fogel, A.: The temporal structure of face-to-face communication between mothers and infants. Dev. Psychol. **16**(5), 454–464 (1980)
Koterba, E.A., Iverson, J.M.: Investigating motionese: the effect of infant-directed action on infants' attention and object exploration. Infant Behav. Dev. **32**(4), 437–444 (2009)
Kuhl, P.K., Andruski, J.E., Chistovich, I.A., Chistovich, L.A., Kozhevnikova, E.V., Ryskina, V.L., Stolyarova, E.I., Sundberg, U., Lacerda, F.: Cross-language analysis of phonetic units in language addressed to infants. Science **277**(5326), 684–686 (1997)
Marshall, P.J., Young, T., Meltzoff, A.N.: Neural correlates of action observation and execution in 14-month-old infants: an event-related EEG desynchronization study. Dev. Sci. **14**(3), 474–480 (2011)
Moore, C., Dunham, P.J. (eds.): Joint Attention: Its Origins and Role in Development. Lawrence Erlbaum, Englewood Cliffs (1995)
Nagai, Y., Asada, M.: Predictive learning of sensorimotor information as a key for cognitive development. In: Proceedings of the IROS2015 Workshop on Sensorimotor Contingensies for Robotics, Hamburg (2015)
Nagai, Y., Rohlfing, K.J.: Computational analysis of Motionese toward scaffolding robot action learning. IEEE Trans. Auton. Ment. Dev. **1**(1), 44–54 (2009)
Nagai, Y., Hosoda, K., Morita, A., Asada, M.: A constructive model for the development of joint attention. Connect. Sci. **15**(4), 211–229 (2003)
Nagai, Y., Asada, M., Hosoda, K.: Learning for joint attention helped by functional development. Adv. Robot. **20**(10), 1165–1181 (2006)
Nagai, Y., Kawai, Y., Asada, M.: Emergence of mirror neuron system: immature vision leads to self/other correspondence. In: Proceedings of the 1st Joint IEEE International Conference on Development and Learning and on Epigenetic Robotics, Frankfurt (2011)
Nagai, Y., Nakatani, A., Qin, S., Fukuyama, H., Myowa-Yamakoshi, M., Asada, M.: Co-development of information transfer within and between infant and caregiver. In: Proceedings of the 2nd IEEE International Conference on Development and Learning and on Epigenetic Robotics, San Diego (2012)
Piaget, J.: The Origins of Intelligence in Children. International Universities Press, Madison (1952)
Rizzolatti, G., Sinigaglia, C.: Mirrors in the Brain: How Our Minds Share Actions and Emotions. Oxford University Press, Oxford (2008)
Rizzolatti, G., Fogassi, L., Gallese, V.: Neurophysiological mechanisms underlying the understanding and imitation of action. Nat. Rev. Neurosci. **2**, 661–670 (2001)
Rohlfing, K.J., Fritsch, J., Wrede, B., Jungmann, T.: How can multimodal cues from child-directed interaction reduce learning complexity in robots? Adv. Robot. **20**(10), 1183–1199 (2006)
Scaife, M., Bruner, J.: The capacity for joint visual attention in the infant. Nature **253**, 265–266 (1975)
Schreiber, T.: Measuring information transfer. Phys. Rev. Lett. **85**(2), 461–464 (2000)

Shibata, M., Fuchino, Y., Naoi, N., Kohno, S., Kawai, M., Okanoya, K., Myowa-Yamakoshi, M.: Broad cortical activation in response to tactile stimulation in newborns. Neuroreport **23**(6), 373–377 (2012)

Umiltà, M.A., Kohler, E., Gallese, V., Fogassi, L., Fadiga, L., Keysers, C., Rizzolatti, G.: I know what you are doing. A neurophysiological study. Neuron **31**(1), 155–165 (2001)

Vygotsky, L.S.: Interaction between learning and development. In: Mind in Society: The Development of Higher Psychological Processes, pp. 79–91. Harvard University Press, Massachusetts (1978)

Zukow-Goldring, P., Arbib, M.A.: Affordances, effectivities, and assisted imitation: caregivers and the directing of attention. Neurocomputing **70**(13–15), 2181–2193 (2007)

Chapter 4
Mirror Neuron System and Social Cognitive Development

Minoru Asada

Abstract This chapter, first, reviews the properties and potentials of the mirror neuron system (MNS) for understanding other's actions and imitation. Next, the developmental process from the emergence of the MNS to social interaction is investigated from a viewpoint of constructive approach. Lastly, in the concluding remarks, future issues are discussed.

Keywords Mirror neuron system (MNS) • Cognitive developmental robotics (CDR) • Empathy • Hebbian learning • Self-organizing mapping (SOM) • Synchronization • Ecological self • Interpersonal self • Social self • Imitation

4.1 Introduction

The mirror neuron system (MNS) is activated when executing a specific action or when observing the same action by other individuals (Rizzolatti et al. 2008). Many researchers in neuroscience, developmental psychology, and even robotics have been investigating the properties of the MNS: e.g., how it represents self/other correspondence and how it leads to social behaviors. Notable findings, among others, are those of the ability to understand other's actions (e.g., Pellegrino et al. 1992; Umilta et al. 2001) and imitation (e.g., Heiser et al. 2003; Nishitani and Hari 2000). However, despite many findings, the origin of the MNS remains a mystery.

In this chapter, we first review the properties and potentials of the MNS for understanding other's actions and imitation. Several studies are introduced and discussed with regard to the design of MNS structure and related functions. Next, we discuss the developmental process from the emergence of the MNS to social interaction from a viewpoint of constructive approach based on cognitive developmental robotics (Asada et al. 2009) (hereafter, 'CDR' in short). Fetal simulation is introduced to show the importance of physical embodiment and interaction at the early stage of life. In addition, the topic of learning of facial structure with

M. Asada (✉)
Graduate School of Engineering, Osaka University, Toyonaka, Osaka, Japan
e-mail: asada@ams.eng.osaka-u.ac.jp

M. Kasaki et al. (eds.), *Cognitive Neuroscience Robotics A*,
DOI 10.1007/978-4-431-54595-8_4

weak vision and tactile sensation is briefly introduced. As for social interaction, it is shown that the vowel acquisition of an infant robot in vowel imitation is affected by caregiver's affirmative biases. Furthermore, intuitive parenting is utilized to develop emotional states of artificial infant. Finally, in the concluding remarks, future issues are discussed.

4.2 MNS and Infrastructure for Social Cognitive Development

After Gallese et al. (1996) found mirror neurons in the ventral premotor cortex of the macaque monkey brain, plenty of evidence has been found in different studies (e.g., Rizzolatti et al. 2008; Ishida et al. 2010). These are:

1. A mirror neuron fires when a monkey acts or when a monkey observes the same action performed by another.
2. As long as the goal is shown, a mirror neuron is activated even if the intermediate trajectory is invisible. It is activated only when action is transitive and has an object as a target of action.
3. A mirror neuron of a monkey reacts to sounds caused by action; it is activated when the monkey does the same action. It is considered that it attends to other's action understanding.
4. Some mirror neurons respond when a monkey observes tool-use by others. Even if the goal is the same, its realization may differ.
5. Some mirror neurons respond to oral communication between monkeys.
6. Mirror neurons exist not only in the ventral premotor cortex but also in the PFG area[1] (neuron activities responding to somatosensory and visual stimuli) at the inferior parietal lobule which is anatomically connected to the ventral premotor cortex.
7. Responses of mirror neurons in the PFG area differ depending on action executors.
8. Visual neurons at the peripheral area of the superior temporal sulcus (STS) of the temporal lobule respond to others' actions. However, they are not called mirror neurons, since they do not have any responses related to motions. Their responses differ depending on gaze, and therefore, a relation between them and joint attention is suggested.
9. The ventral premotor cortex F5, the PFG area at the inferior parietal lobule, and the peripheral area of STS are anatomically connected and constitute a system called the "MNS."
10. Mirror neurons related to reaching are recorded at the dorsal premotor cortex.

[1]PFG area is between the PF and PG areas, and PF, PG, and PFG are vernacular terms for monkey brain regions. Actually, PF and PG correspond to the gyrus supramarginalis (Brodmann areas 7b) and gyrus angularis (7a), respectively.

Since the ventral premotor cortex of the macaque monkey brain may correspond to the human brain regions that are close to the Broca area, and the Broca area is related to language faculty, Rizzolatti and Arbib (1998) suggest a close relationship between the MNS and language acquisition. They also speculate and expect the elucidation of the process in which humans have acquired cognitive functions, such as imitation, action understanding, and further language faculty. Some of these functions are discussed in the book (2006) edited by Arbib.

An unsolved issue is how other person's action is realized in one's own brain. That is, one's capability to simulate other's internal state has not been unraveled (Gallese and Goldman 1998). This issue is related to self/other discrimination, action understanding, joint attention, imitation, theory of mind, empathy, and so on. Ishida et al. (2010) and Murata and Ishida (2007) suggest the following on the basis of the above findings and related ones:

- Coding of other's body parts occurs in the same area that encodes one's own body parts, and the map of one's own body parts is referred to in perception of other's bodies.
- As a step to self body cognition, the coincidence of efference copy and sensory feedback forms the sense of self-motion. In fact, Murata's group has found the neurons in the parietal area related to the integration of information from efference copy and sensory feedback (Murata and Ishida 2007).
- Originally, mirror neurons had visually coded motions and worked as sensory feedback during motion execution regardless of self or other's motions. The current MNS has been constructed after integration with motion information through the evolutionary and developmental processes.
- Mirror neurons are important for the cognition of self and other's bodies and the recognition of other's motion.
- It is unclear whether the parietal area detects the coincidence of efference copy and sensory feedback (consciousness of the "self") or their difference (consciousness of "the other") (although it is suggested that the human parietal area detects the difference).

Shimada (2009) proposed a model for the activity of mirror neurons with two terms: externalized body representation (mainly based on vision) and internalized body representation (proprioception and efference copy) (similar models were proposed in (Ishida et al. 2010; Murata and Ishida 2007)). In the visual area, the externalized body is represented, and its consistency with the internalized body representation is checked. When one is in this process, "it is not supposed that one is aware of the difference when she discriminates herself from others, but that the internalized body representation is adjusted so that this difference can be canceled, and as a result the motor and sensory areas are activated" (Shimada 2009, p. 72). The processing flow from the externalized body representation to the internalized one seems sufficiently possible, considering that the movement is always adjusted on the basis of visual feedback. The model points out that the integration process of different senses (vision, tactile, auditory, somatosensory, and motor commands) related to the externalized and internalized body representation is shared by self/other discrimination and the MNS.

The process of sharing and discriminating the internal states of both self and others does not seem limited to motor experiences but works on other modalities. Examples may back up this claim: the sensation of being touched when observing others touched by someone (the somatosensory and parietal association areas are activated), pleasure or unpleasure when observing a person smelling (an emotional circuit responds), and the feeling of pain when observing others' pain directly. These examples are supposed origins of empathy. The fact that experiences of the self and others are represented in the common area in the brain can be interpreted to suggest the existence of the mechanism that processes experiences of others in the same way as does those of the self, and such a "mirror-like" property is supposed to be a neural infrastructure for the capability of sharing and understanding other's internal states, including emotion (Fukushima and Hiraki 2009).

Thus, the MNS contributes to the learning and development of social behavior through awareness of the self and others on the basis of the commonality and difference between them. However, it does not seem clear to what extent the MNS is innate and how much is learned through experience. Mirror neurons in monkeys only respond to goals with visible targets (actions of transitive verbs), while the MNS in humans also responds to actions of intransitive verbs without any target (Rizzolatti et al. 2008). How is this difference to be interpreted?

One plausible interpretation is this: in the case of monkeys, goal-oriented behavior needs to be established and used early, due to the high pressure for survival; on the other hand, in the case of humans, this pressure is reduced by caregivers, and therefore, the MNS works not only for goal-oriented behavior but also for behavior without goals. Consequently, much room for learning and structuring for generalization is left, and therefore, more social behaviors and higher cognitive capabilities are expected to be acquired by learning. Shimada (2009) mentions that "it is not supposed that one is aware of the difference when she discriminates herself from others, but that the internalized body representation is adjusted so that the difference can be canceled, and as a result the motor and sensory areas are activated" (Shimada 2009, p. 72). If this is correct, repeating such behavior is supposed to be linked to imitation and communication. In the case of monkeys, the motivation to cancel this difference seems low due to the high pressure for survival and therefore no link to imitation. Actually, monkeys are known not to have the ability to imitate.

4.3 Sociality Developmental Model by CDR

Explanation and design of the developmental process from fetus to child are given by a synthetic approach. It involves a viewpoint of sociality development. The basic outlines of our synthetic approach are as follows:

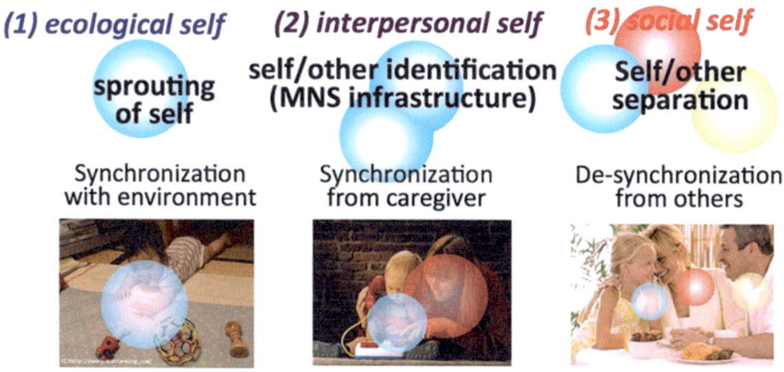

Fig. 4.1 The developmental process of establishing the concept of self and other(s) (Asada 2015)

1. Since we argue on the level of not phylogeny but ontogeny, we minimize the embedded structure, and maximize the explanation and design, of the learning process. Body structures, especially brain regions and sensory organs, configuration of muscle-skeleton system, and their connections, are embedded structures. They change in the process of development. Other developmental changes (e.g., connectivity and synaptic efficacy) are considered at each stage of sociality development.
2. The axis of sociality is formed according to the level of self/other cognition. Changes from the implicit others (including the environment) who can be perceived as nonself to the explicit others who are similar to the self constitute the fundamental axis of development.
3. During this process, the role of the MNS should be made clear, and we discuss any possibilities of its construction and use.
4. We use Hebbian learning and self-organizing mapping (SOM) as learning methods. In fact, most of the synthetic approaches proposed so far have used them.

Figure 4.1 shows the developmental process of establishing the concepts of self and other(s), partially following Neisser's definition of "self" (Neisser 1993). The term "synchronization" is used as a keyword to explain how these concepts develop through interaction with the external world including other agents. We suppose that there exist three stages of self-development and they are connected seamlessly.

The first stage is the period in which an agent forms the most fundamental concept of self through physical interaction with objects in the environment. At this stage, synchronization with objects (more generally, environments) through rhythmic motions, such as beating, hitting, knocking, and so on, or reaching behavior is observed. Synchronization tuning and prediction are the main activities of the agent. If completely synchronized, the phase is locked (phase difference is zero), and both the agent and the object are mutually entrained into the synchronized state. In this phase, we may say that the agent has its own representation of the so-called ecological self; the term stems from Gibsonian psychology, which claims that infants can receive information directly from the environment, for their sensory organs are tuned to certain types of structural regularities (Hardcastle 1995). Neural oscillation might be a strong candidate for the fundamental mechanism for enabling such synchronization.

The second stage is the period in which self/other discrimination starts with supports from the MNS infrastructure inside and caregivers' scaffolding outside. During the early period in this stage, infants regard caregiver's actions as their own (the "like me" hypothesis Meltzoff 2007) since a caregiver works as one of others who can synchronize with the agent. The caregiver helps the agent consciously and sometimes unconsciously in various manners, such as motherese (Kuhl et al. 1997) or motionese (Nagai and Rohlfing 2009). This synchronization may be achieved through turn-taking, which includes catching and throwing a ball, giving and taking an object between the caregiver and agent, or calling each other. Then, infants gradually discriminate a caregiver's actions as other's (the "different from me" hypothesis, Inui 2013). This is partially because caregivers help infants' actions first and then gradually promote their own action controls (less help), and partially because not-yet-matured sensory and motor systems of infants make it difficult to discriminate between self's and other's actions at the early period in this stage. In the later period in this second stage, an explicit representation of others emerges in the agent, while no explicit representation of others is in the first stage, even when caregivers interact with the agent. The phase difference in turn-taking is supposed to be 180°. Due to the explicit representation of others, the agent may have its own self representation of the so-called interpersonal self. At the later stage of this phase, the agent is expected to learn when she should inhibit its behavior by detecting the phase difference so that turn-taking between the caregiver and self can occur.

This learning is extended in two ways. One is recognition, assignment, and switching of roles, such as throwing and catching, giving and taking, and calling and hearing. The other is learning to desynchronize from the synchronized state with a person and to start synchronization with another person when the agent has some reason to do so, such as sudden leave of a close person (passive mode) or attention to some person (active mode). Especially, the latter needs active control of synchronization (switching), and this active control facilitates the agent to take a virtual role in make-believe play. At this stage, the targets to synchronize include not only persons but also objects. However, the synchronization with objects is not the same as the first stage in that the target includes virtualized but unreal objects, such as virtualized mobile phone, virtualized food in the make-believe play of eating or giving, and so on. If some behavior relevant to this stage is observed, we can say that the agent has the representation of the "social self." In what follows, we review several studies concerning the processes of these three stages.

4.3.1 Self/Other Undifferentiated State

Recent progress in visualization technology, such as 4-D sonar, enables us to reveal various fetal behaviors and capabilities in the womb (e.g., Myowa-Yamakoshi and

Takeshita 2006). Stimuli from inside and outside the mother's womb, such as lights and sounds, are implicit others and interact with the fetus. The fetus is yet incapable of differentiating self from others.

The onset of the first sensation, touch, is supposed to start 10 weeks after conception; and vision is supposed to start 18–22 weeks after conception (http://www.birthpsychology.com/lifebefore/fetalsense.html). Assuming that body representation can be obtained on the basis of cross-modal representation, it is easy to hypothesize that self body representation can be acquired to some extent from the motor and by somatosensory learning before perceiving others' bodies visually.

The visual and auditory systems begin to work around 20 weeks after conception. Their connection to vocalization and limb movements does not seem strong, and therefore, they are undifferentiated and immature. However, coordinated motions between lip and hand, such as thumb-sucking, seem innate or already learned because of body posture constraints in the womb (e.g., a hand approaching to the lip when the mouth opens (Myowa-Yamakoshi and Takeshita 2006)). In this period, the individual motor library starts to be acquired as the infrastructure of the MNS.

Kuniyoshi and Sangawa (2006) constructed a simulated model of a fetus and showed that various meaningful motor patterns emerged after the birth without "innate" motor primitives. The model consists of a musculoskeletal body floating in a uterus environment (elastic wall and liquid) and a minimal nervous system having spine, medulla, and primary sensory/motor cortical areas; its connection weights between brain regions are initially random. Hebbian learning and self-organizing mapping methods are used to determine the connection weights. After learning, the configuration of the muscle units and more generally the cortex map for somatosensory and motor areas are acquired, and the fetal movements in the womb change from random to ordered ones. In the simulation after the birth with gravity, the "neonate" model exhibits emergent motor behaviors, such as rolling over and crawling-like motion. These observed whole-body motor patterns are purely emergent from the interaction between the body, environment, and nervous system. This is a typical example that "body shapes brain" (Kuniyoshi 2008). This approach can be regarded as a principle of constructive approaches for individual development. Recently, Kuniyoshi's group increased the granularity of their simulation for the brain, body, and environment, to elucidate the principles of social behavior. For example, Mori and Kuniyoshi (2010) compared how different tactile distributions affect the developmental process. They tested a heterogenous tactile distribution (similar to the tactile distribution of real fetus) and a uniform distribution (not realistic but virtual). Figure 4.2a shows an example of hand-face contacts. Each circle represents a tactile point. Red color represents outputs of mechanoreceptors corresponding to tactile points. The body is transparent. Figure 4.2b, c shows hand-face contacts in cases of uniform (not realistic) tactile and heterogenous distributions, respectively. We can see that the former does not change so much, while the latter gradually increases and resembles a real fetus's behavior.

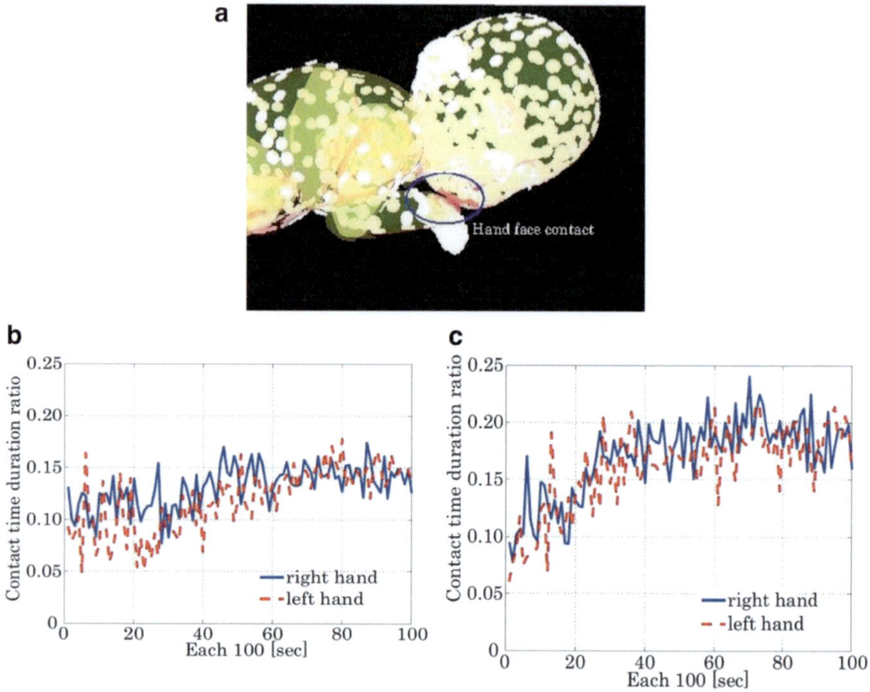

Fig. 4.2 Comparison of hand-face contacts in terms of the difference of tactile distribution (From Mori and Kuniyoshi 2010). (**a**) An example of hand and face contact. (**b**) Hand and face contacts with the uniform tactile distribution. (**c**) Hand and face contacts with the realistic tactile distribution

4.3.2 The Beginning of Discrimination Between Self and Nonself

Neonatal imitation was found in Meltzoff and Moore (1977). It has been a hot topic and caused a controversy concerning what is "innate" and what is "learned." As seen from the 4-D ultrasound imaging of fetus movements, fetuses start touch motions with their body parts, such as the face and arm, at least 14 or 15 weeks after gestation. Hand movements toward the lip are often observed, and these body parts have a high-density distribution of tactile sensor organs. Can these touch motions provide clues for the "innate" or "learned" nature of neonatal imitation?

We propose two main causes of these touch motions from the viewpoint of CDR: a physical constraint on the body structure (posture constraint) and information maximization tendency. As for the former, a fetus folds her arms much more often than she extends them inside the tight space of the womb. Thus, there is a high probability of the hands being positioned close to the face, in particular to the mouth. In this case, little freedom of hand-arm movements is given to the fetus because of the limitations of not only the joint structure but also the muscle attachment layout. Information maximization tendency, that is, touching the lip with a hand, may allow for acquiring more information than touching other body parts. This is because the hands and lip have a high-density distribution of tactile sensory organs. Furthermore, the oral cavity, differing from the body surface, is an interesting target for exploratory behavior.

In the fetal and neonatal periods, self sensory/motor mapping (especially, between hand movements close to the mouth and lip) is acquired. This may be the period in which the concept of others has not been matured, but the concepts of ecological self and nonself begin to be discriminated. The fundamental issue is self body cognition. With the contingency principle of self body cognition in place, Miyazaki and Hiraki (2006) argued, using delays of visual sensation, that infants showed a cognitive sensitivity of self body cognition. Asada et al. (1999) proposed a method for estimating the spatial and temporal dimensions for visual stimuli caused by the self-induced motions, as a computational model of self/other discrimination. The method can extract a static environment, self body, or stationary objects, in synchronization with the self body. The method is based on the fact that (body) observation is directly related to self-induced motion. In this case, the temporal dimension is one (the spatial dimension depends on the complexity of the object structure). The temporal dimension greater than one includes others or moving objects manipulated by them. However, there is no exact representation of others similar to the self.

4.3.3 *The Existence of Explicit Others*

After birth, infants gradually develop body representations, categories for graspable objects, mental simulation capabilities, and so on, through learning processes. For example, hand regard at the fifth month is the learning of the forward and inverse models of the hand. Table 4.1 shows typical behaviors and their learning targets in infant's learning processes. Infants establish the ecological self and nonself and

Table 4.1 Infant development and learning targets (Asada et al. 2009)

Months behaviors		1.1 Learning targets
5	Hand regard	Forward and inverse models of the hand
6	Finger the other's face	Integration of visuo-tactile sensation of the face
	Observe objects from different viewpoints	3-D object recognition
7	1.1.1 Drop objects and observe the result	1.1.2 Causality and permanency of objects
8	Hit objects	Dynamics model of objects
9	Drum or bring a cup to the mouth	Tool use
10	Imitate movements	Imitation of invisible movements
11	Fine grasp and carry objects to others	Action recognition and generation, cooperation
12	Pretend	Mental simulation

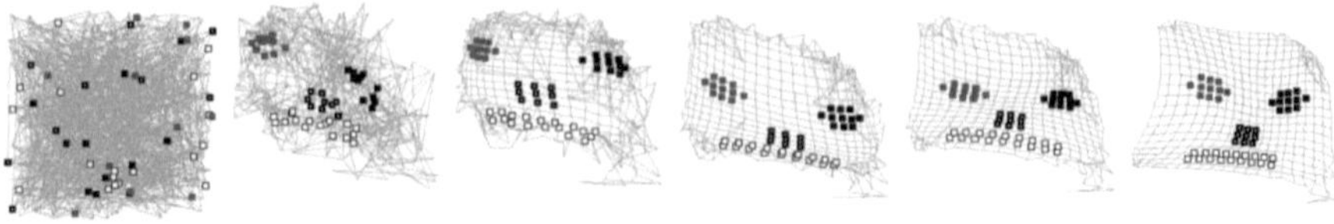

Fig. 4.3 The configuration of the tactile sensor units from the random state (the *leftmost*: the 1st step) to the final one (the *rightmost*: the 7200th step), where *gray* squares, *black* ones, empty squares with *thin lines*, and empty squares with *thick lines*, respectively, correspond to right eye, left one, mouth, and nose (From Fuke et al. 2007)

also acquire comprehension of the interpersonal self, explicit others, and a trinomial relationship (self, other, and an object).

In sociology, the cognition that the self and others are similar is supposed to be a precondition for estimating the internal states of others. In accordance with preceding studies, Sanefuji and Ohgami (2009) argued that infants were able to discover the similarity of themselves in someone else's body or behavior on the basis of understanding of self/other similarities. In addition infants can relate changes in other's mind that are caused by the behaviors. That is, understanding of self/other similarity is supposed to be an infrastructure which supports one aspect of social cognitive developments, such as understanding of the other's demands, intention, and emotion (Meltzoff 2007).

If seen from the viewpoint of CDR, the basic requirements for similarity understanding are the localization of self facial parts, facial pattern detection in visual observation of others, and the correspondence of facial parts between the self and others. Fuke et al. (2007) proposed a possible learning model that enabled a robot to acquire a body image of its body parts even when they were invisible to itself (in this case, that was its "face"). The model associates spatial perception based on motor experience and motor images with perception based on the activations of touch sensors and tactile image both of which are supported by visual information. Moreover, Fuke et al. (2007) proposed a method to detect facial parts, such as eyes, nose, and mouth, each of which has a unique tactile pattern observed by hand touching and scanning on its face. The information on these facial parts, once detected, can be used to construct a correspondence of face between the self and the other. Figure 4.3 shows the configuration of the tactile sensor units from the random state to the final one, where gray squares, black ones, empty squares with thin lines, and empty squares with thick lines correspond to the right eye, left one, mouth, and nose, respectively.

Understanding of face similarity between oneself and others is linked with understanding of the similarity between one's body and other's and moreover with understanding of the other's behaviors. Before turning to the next issue concerning understanding of other's behaviors, we need to solve the coordinate transformation problem between oneself and others. The problem consists in absorbing the apparent difference of the same motions performed by different agents (self or other). Ogawa and Inui (2007) suggest that the parietal area performs coordinate transformation between the egocentric and allocentric bases. From the viewpoint of CDR, we would like to search for any possibility of learning of coordinate transformation after birth and not to suppose that it is innate. How is the learning of coordinate transformation possible?

One possibility is suggested by the scheme of reinforcement learning in which a value is assigned to each state according to the distance to the goal, regardless of the difference between apparent motions. If we know the goal and can measure the distance to it from the current state, the different views (states) formed by observations from different coordinate systems can be regarded as identical if their values are the same (Takahashi et al. 2010).

States with the same value can be regarded as the same from a viewpoint of goal achievement, and apparent differences among them are supposed to stem from differences of coordinate systems. Since the purpose of coordinate transformation is to equalize the states beyond apparent differences, coordinate transformation can be realized if the agent knows the values of all states (views), including the observation of self and other's behavior.

4.3.4 *Interaction Between the Self and a Caregiver as the Other*

The concept of others, and in particular the concept of caregivers who have an exact role, is established during the period of infancy, in which the conceptual self is also established. Typical issues are the emergence of the MNS, joint attention, imitation, and the development of empathy (see also, Asada 2015) . In what follows, we briefly deal with these issues.

4.3.4.1 Emergence of MNS

One big question is how the mirror neuron system (MNS) develops, and this question has attracted increased attention of researchers. Among various hypotheses, a widely accepted model is associative sequence learning, in which the MNS is acquired as a byproduct of sensorimotor learning (e.g., Ray and Heyes 2011). The model, however, cannot discriminate the self from others since it adopts too simplified sensory representations.

Nagai et al. (2011) have proposed a computational model for the early development of the self/other cognition system which originates from immature vision. The model gradually increases the spatiotemporal resolution of a robot's vision as the robot learns sensorimotor mapping through primal interactions with others. In the early stage of development, the robot interprets all observed actions as equivalent because of its low visual resolution. The robot thus associates non-differentiated observations with motor commands. As vision develops, the robot starts discriminating actions generated by itself from those by others. The initially acquired association is, however, maintained through development and results in two types of associations: association between motor commands and self-observation and association between motor commands and other observation. Their experiments demonstrate that the model achieves the early development of the self/other cognition system and enables a robot to imitate others' actions (the next chapter will show the details).

4.3.4.2 Joint Attention and Imitation

Existing approaches to joint attention have mainly dealt with gaze control (e.g., Nagai et al. 2003, 2006), and it seems difficult for them to satisfy the statement proposed by Emery et al. (1997): "Joint attention has the additional requirement that X follows the direction of Y's gaze to the object (Z) that is the focus of Y's attention" (Emery et al. 1997, p. 286). Any scheme for self/other discrimination and recognition of other's behavior is supposed to enable an agent to develop gaze control: from gaze control as a fundamental behavior of joint attention to joint visual attention as purposeful behavior to follow the other's gaze by inferring the target that the other attends (Butterworth and Jarrett 1991). During these processes, the agent's attention is represented inside the self; and then it is shared in the way that the self's and the other person's attention are the same (Shimada 2009). This can be regarded as exactly what the MNS does. Through these processes, the agent is expected to acquire certain social behaviors, such as social referencing (a kind of gaze behavior designed to awaken the other's attention).

Regardless of its explicit or implicit representation, coordinate transformation between the self and the other may aid imitation significantly. The corresponding issue in vocal imitation and especially in vowel imitation is the transformation of each vowel between the self and the other in the formant space; this kind of transformation often happens between an infant and her caregiver. A consequence of this correspondence is to share each symbol (vowel), such as //a//, //e//, //i//, //o//, or //u//, between them. A synthetic approach assumes a strong bias that a caregiver provides goal-oriented information, regardless of its explicit or implicit representation. This is confirmed by Yoshikawa et al. (2003). They show that mutual imitation accelerates the learning of vowel imitation. They hypothesize that preference for imitation itself is a precondition for this learning. Again, the adjusting structure suggested by Shimada (2009) may be linked to this preference. With a similar experimental setting to Yoshikawa et al. (2003), Miura et al. (Butterworth and Jarrett 1991) have argued that being imitated by a caregiver has two meanings: not only providing information on the correspondence of vowels between the robot and the caregiver but also guiding the robot to perform what the caregiver does in a more similar way.

Being inspired by the preceding work (Miura et al. 2007), Ishihara et al. (2009) have computationally modeled an imitation mechanism as a Gaussian mixture network (GMN), some parameters of which are used to represent caregiver's sensorimotor biases, such as the perceptual magnet effect (Kuhl 1991); the effect

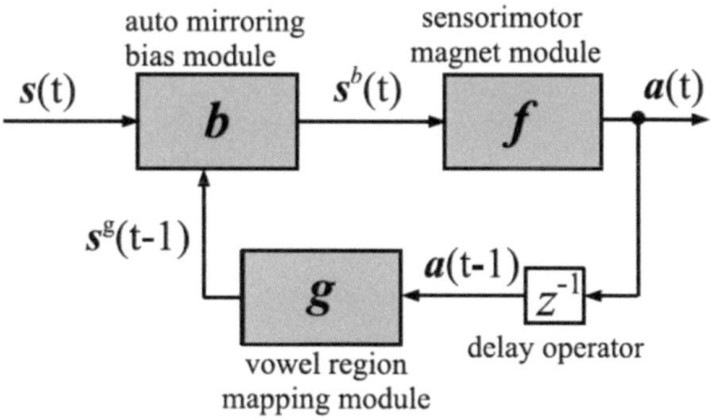

Fig. 4.4 Vocal imitation mechanism (From Ishihara et al. 2009)

is used in the computer simulation to evolve a population to share vowels (Oudeyer 2002). The perceptual magnet effect indicates a psychological phenomenon where a person recognizes phonemes as more typical in the phoneme categories that the person possesses in her mind than they actually are. Ishihara et al. have conducted a computer simulation where an infant and a caregiver imitate each other with their own GMN; the GMN of the infant is learnable and that of the caregiver involves a certain level of magnet effects. The result of the simulation is that the caregiver's imitation with the magnet effects can guide the infant's vowel categories toward corresponding ones. Interestingly, the effectiveness of guidance is enhanced if there is what is called the "automirroring bias" in the caregiver's perception, that is, if she perceives the infant's voice as closer to the one that resembles her precedent utterance than to the real one.

Figure 4.4 shows the vocal imitation mechanism proposed by Ishihara et al. (2009); the mechanism involves the automirroring bias and the sensorimotor magnet effect as two modules. Figure 4.5a shows the learning result with two well-balanced biases, where blue and red dots indicate the caregiver's and the infant's voices, respectively; apexes of red pentagons represent target vowels of the infant, i.e., clearest vowels in her vowel region; and black dots represent infant vowel prototypes after learning. In (a), the infant vowels are almost correctly converged, while, in (b), (c), and (d), they do not seem to be. In (b), the sensorimotor magnet is missing. Therefore, a divergence from the infant's desired prototypes is observed, and the caregiver imitates this divergence. On the other hand, in (c), the automirroring biases are missing, and as a result, there is a fast convergence to the prototypes; but three of the five vowels have converged in wrong locations. In the case of no biases, nothing has happened as shown in (d).

Although previous synthetic studies focus on finding a correspondence between infant's and caregiver's vowels formed by imitation, they have all assumed for simplicity that infants are almost always or always imitated by caregivers. This is apparently unrealistic. Realistically, infants usually become able to realize that they are being imitated. Miura et al. (2012) have addressed this issue by considering the low rate of being imitated in computer simulation and proposed a method called "autoregulation": it is the active selection of action and data with underdeveloped classifiers of caregiver's imitation of infant's utterances.

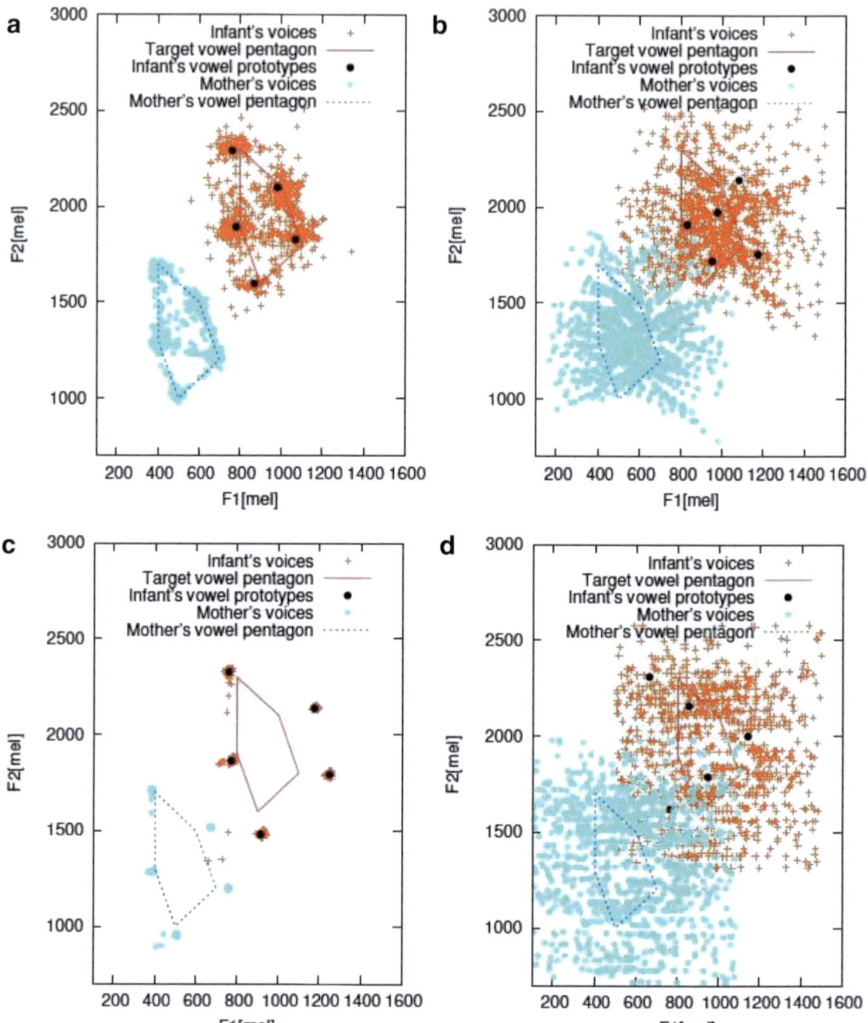

Fig. 4.5 Different learning results under several conditions. Apexes of *red* pentagons represent target vowels of infant, i.e., clearest vowels in her vowel region, and *black dots* represent infant vowel prototypes after learning (From Ishihara et al. 2009). (**a**) Both biasing elements. (**b**) Only automirroring bias. (**c**) Only sensorimotor magnets. (**d**) No biasing element

4.3.4.3 Empathy Development

In the previous section, the importance of "face" was pointed out as a starting point for understanding the similarity between the self and others. A technical issue is how to improve face recognition and understanding of facial expressions. Behavior generation based on recognition of other's facial expressions (including detection of her gaze) is extremely important for communication. Therefore, CDR should overcome the difficulty in designing and building artificial faces.

Based on assumption of intuitive parenting, Papousek and Papousek (1987) and Watanabe et al. (2007) built a face robot that empathizes with its caregiver. A typical example of intuitive parenting is that a caregiver mimics or exaggerates a child's emotional expressions. This is regarded as a good opportunity to teach children how to feel in real time (most adults have a skill for this). Children are able to understand the meaning of facial expressions and develop empathy toward others when the process is reinforced with an emphasis on the facial expressions of their caregivers. This is because children empirically learn the connections between their internal states and the facial expressions of others.

Figure 4.6a shows a learning model for a child developing a sense of empathy through the intuitive parenting of her caregiver. When a child undergoes an emotional experience and expresses her feelings by changing her facial expression, the caregiver sympathizes with the child and shows a concomitantly exaggerated facial expression. The child then discovers the relationship between the experienced emotion and the caregiver's facial expression and associates them (Fig. 4.6b). The emotion space in this figure is constructed on the basis of the model proposed by Russell (1980).

Since this study focuses on the association between infant's internal state and her caregiver's facial expressions, there is no self/other discrimination in the child's emotional space. In developing the conceptual self, cognition of similar but not identical others leads to the continuous process of filling the gap between the self and others, and this process is linked to imitation and communication.

4.3.5 *Social Development of Self-Concept and MNS*

We have seen the relationship between the social development of the concept of self and the MNS from the viewpoint of synthetic approach and reviewed the issues of CDR. As mentioned above, instead of representing the self independently depending on the cognitive functions during the period of development, a consistent representation of the self (and others) that explains and designs cognitive development is needed to link the emergence of a new paradigm. That is, the development of the concept of self, as shown in Fig. 4.1, is not a sequence of isolated representations of different selves; rather, it is expected to be a consequence of the emergence or development with a consistent structure. Since the seamless development of self representation is so difficult, we cannot expect an immediate and unique solution to it yet. Therefore, we need to divide the whole issue into several parts (some of which may be overlapping) and address each part separately. Attention should be paid to the separation between the innate cognitive functions and those that emerge after learning and development. Moreover, the relationship between them should be focused upon. For example, dynamic changes associated

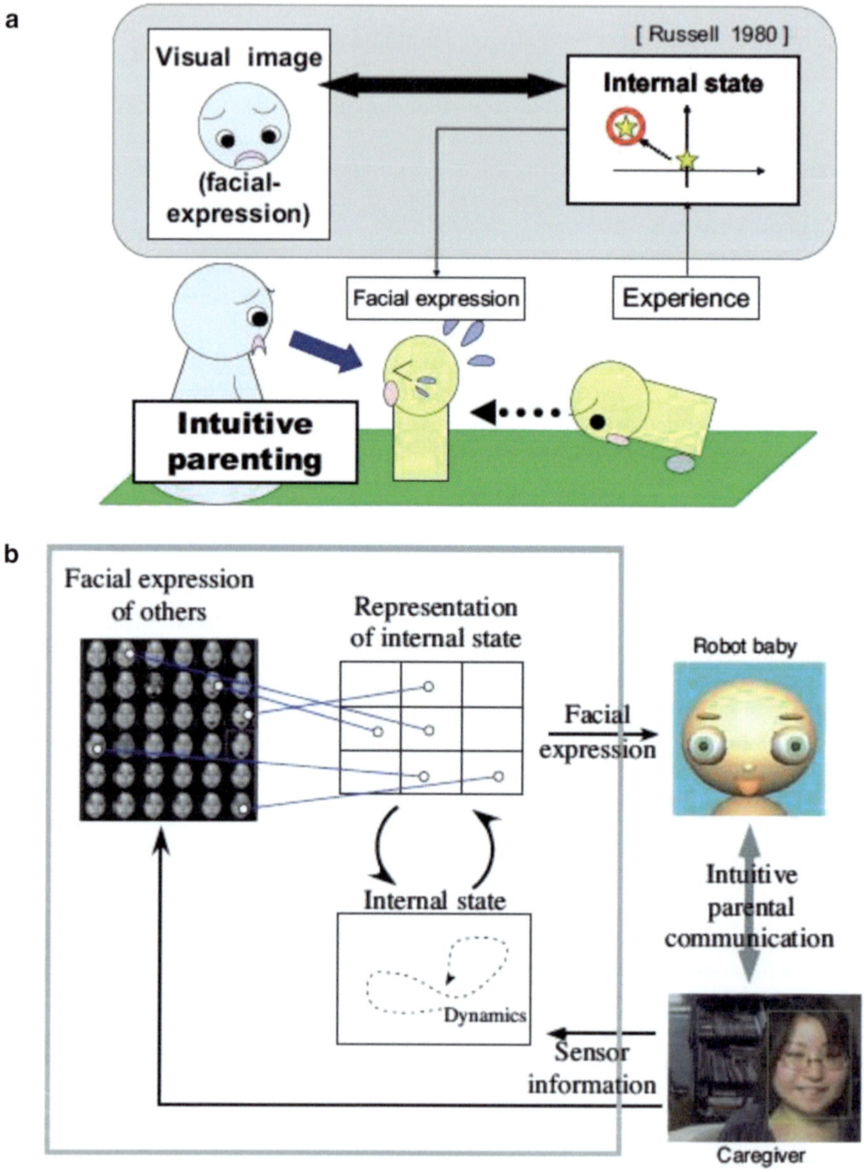

Fig. 4.6 A learning model for a child developing a sense of empathy through the intuitive parenting of its caregiver (From Watanabe et al. 2007). (**a**) A concept of learning model for developing empathy in children through intuitive parenting. (**b**) An architecture which associates visual facial expressions of others with internal states

with independent development, mutual acceleration, and interference are important relationships to study. Through the accumulation of these studies, we may shed light on self-development, which is linked to the creation of a new paradigm.

The solution for the issues of language and theory of mind is a symbolic goal for CDR. During the developing processes of the concepts of the self and others, the development of vocal communication is expected from the action-based communication (Arbib 2006). To cope with these issues, a development of other research platforms is important and necessary. In addition, the issue of memory is essential, although we still need time to solve it. To have the concept of time, robots need to realize the decay of its physical body. An emotional state, e.g., sadness, is originally based on the self body. A body design is needed to handle these issues.

4.4 Discussion

We have reviewed the synthetic approaches in CDR and sought any possibility of designing subjectivity. We then have discussed how the MNS can help the issue of designing subjectivity. However, the MNS itself is a little controversial (Hickok 2009), and we should avoid too much speculation. For this reason, we should focus on an essential mechanism or principle that plays the same role that the MNS is expected to play. Instead of designing the function of the MNS directly, we need to build a more general and fundamental structure. Among several candidates, a promising one is synchronization of oscillations; it is widely observed in neural assemblies in the brain.

Yamaguchi (2008) investigated theta rhythms in rat hippocampus and in human scalp EEG in order to clarify the computational principle in these systems with essential nonlinearity. They claimed that synchronization can crucially contribute to computation through unification among heterogeneous developing systems in real time. Furthermore, Taga's group (Homae et al. 2007; Watanabe et al. 2008; Nakano et al. 2009) found synchronization of brain activities; in particular, a correlation between the frontal regions at the age of 2 or 3 months and one between the frontal and occipital regions at the age of 6 months were found on the basis of the measurement of brain activities by NIRS. One idea would be to introduce a concept of "synchronization" to explain and design the emergence of subjectivity and more correctly the phenomenon that an agent can be regarded as having her subjectivity.

In Sect. 4.3, we have discussed how the concept of self develops through the interactions with the environment including other agents. The developmental process shown in Figs. 4.1 and 4.7 indicates the mechanisms that correspond to these three stages. The common structure is a kind of mechanism of "entrainment," and the target with which the agent harmonizes may change; more correctly, she may be endowed with corresponding substructures. In the first stage, a simple synchronization with objects is realized, and a caregiver initiates the synchronization in the second stage in which a sort of representation of agent (agency) is gradually established and a substructure of inhibition is added for turn-taking.

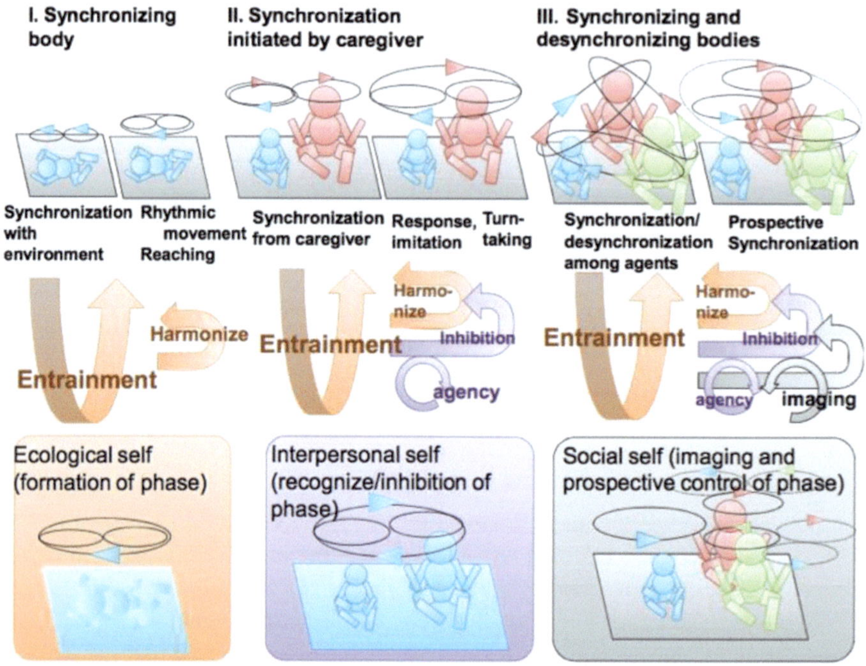

Fig. 4.7 Three mechanisms for the development of "self" (Asada 2015)

Finally, another control skill of synchronization is added to switch the target agents for harmonization. Imaginary actions toward objects could be realized on the basis of a sophisticated skill in switching. These substructures are not added but expected to emerge from the previous stages. A long-ranging research issue is what kind of more fundamental structure enables the emergence of the substructures.

Exercises

Study more about the MNS, and answer the following questions:

1. In what region of the monkey brain were mirror neurons originally found? What regions in the human brain correspond to those important regions in the monkey brain, and what kind of functions are they expected to have?
2. Hebbian learning is explained as "Fire together, wire together," which means that synaptic weights between neurons increase as their neuron firing increases. However, this seems unrealistic since it only implies increases in synaptic weights. Describe and explain a more realistic mechanism for Hebbian learning and other variations to enhance the learning.

3. Self-organizing mapping (SOM) is a standard method for clustering. Explain the original meaning and typical method of SOM.
4. In this chapter, we have discussed how an artificial system could have a kind of structured mind. Describe what you think about such a possibility. If you take the possibility to be very high or low, explain why.

References

Arbib, M.A.: The mirror system hypothesis on the linkage of action and languages. In: Arbib, M.A. (ed.) Action to Language Via the Mirror Neuron System, pp. 3–47. Cambridge University Press, Cambridge/New York (2006)

Asada, M.: Towards artificial empathy. Int. J. Social Robot. **7**(1), 19–33 (2015)

Asada, M., Uchibe, E., Hosoda, K.: Cooperative behavior acquisition for mobile robots in dynamically changing real worlds via vision-based reinforcement learning and development. Artif. Intell. **110**, 275–292 (1999)

Asada, M., Hosoda, K., Kuniyoshi, Y., Ishiguro, H., Inui, T., Yoshikawa, Y., Ogino, M., Yoshida, C.: Cognitive developmental robotics: a survey. IEEE Trans. Auton. Ment. Dev. **1**(1), 12–34 (2009)

Butterworth, G.E., Jarrett, N.L.M.: What minds have in common is space: spatial mechanisms serving joint visual attention in infancy. Br. J. Dev. Psychol. **9**, 55–72 (1991)

Emery, N.J., Lorincz, E.N., Perrett, D.I., Oram, M.W.: Gaze following and joint attention in rhesus monkeys (*Macaca mulatta*). J. Comp. Psychol. **111**, 286–293 (1997)

Fuke, S., Ogino, M., Asada, M.: Body image constructed from motor and tactile images with visual information. Int. J. Humanoid Rob. **4**, 347–364 (2007)

Fukushima, H., Hiraki, K.: Whose loss is it? Human electrophysiological correlates of nonself reward processing. Soc. Neurosci. **4**(3), 261–275 (2009)

Gallese, V., Goldman, A.: Mirror neurons and the simulation theory of mindreading. Trends Cogn. Sci. **2**, 493–501 (1998)

Gallese, V., Fadiga, L., Fogassi, L., Rizzolatti, G.: Action recognition in the premotor cortex. Brain **119**(2), 593–609 (1996)

Hardcastle, V.G.: In: Kessel, F.S., Cole, P.M., Johnson, D.L. (eds.) A Self Divided: A Review of Self and Consciousness: Multiple Perspectives. Psyche **2**(1) (1995)

Heiser, M., Iacoboni, M., Maeda, F., Marcus, J., Mazziotta, J.C.: The essential role of Broca's area in imitation. Eur. J. Neurosci. **17**, 1123–1128 (2003)

Hickok, G.: Eight problems for the mirror neuron theory of action understanding in monkeys and humans. J. Cogn. Neurosci. **21**, 1229–1243 (2009)

Homae, F., Watanabe, H., Nakano, T., Taga, G.: Prosodic processing in the developing brain. Neurosci. Res. **59**, 29–39 (2007)

Inui, T.: Embodied cognition and autism spectrum disorder (in Japanese). Jpn. J. Occup. Ther. **47**(9), 984–987 (2013)

Ishida, H., Nakajima, K., Inase, M., Murata, A.: Shared mapping of own and others' bodies in visuotactile bimodal area of monkey parietal cortex. J. Cogn. Neurosci. **22**(1), 83–96 (2010)

Ishihara, H., Yoshikawa, Y., Miura, K., Asada, M.: How caregiver's anticipation shapes infant's vowel through mutual imitation. IEEE Trans. Auton. Ment. Dev. **1**(4), 217–225 (2009)

Kuhl, P.K.: Human adults and human infants show a "perceptual magnet effect" for the prototypes of speech categories, monkeys do not. Percept. Psychophys. **50**, 93–107 (1991)

Kuhl, P., Andruski, J., Chistovich, I., Chistovich, L., Kozhevnikova, E., Ryskina, V., Stolyarova, E., Sundberg, U., Lacerda, F.: Cross-language analysis of phonetic units in language addressed to infants. Science **277**, 684–686 (1997)

Kuniyoshi, Y:. Body shapes brain -emergence and development of behavior and mind from embodied interaction dynamics. In: The 10th International Conference on the Simulation of Adaptive Behavior, 2008 (SAB08) Osaka, July 2008, page (an invited talk) (2008)

Kuniyoshi, Y., Sangawa, S.: Early motor development from partially ordered neural-body dynamics: experiments with a cortico-spinal-musculo-skeletal model. Biol. Cybern. **95**, 589–605 (2006)

Meltzoff, A.N.: The 'like me' framework for recognizing and becoming an intentional agent. Acta Psychol. (Amst) **124**, 26–43 (2007)

Meltzoff, A.N., Moore, M.K.: Imitation of facial and manual gestures by human neonates. Science **198**, 74–78 (1977)

Miura, K., Yoshikawa, Y., Asada, M.: Unconscious anchoring in maternal imitation that helps finding the correspondence of caregiver's vowel categories. Adv. Robot. **21**, 1583–1600 (2007)

Miura, K., Yoshikawa, Y., Asada, M.: Vowel acquisition based on an automirroring bias with a less imitative caregiver. Adv. Robot. **26**, 23–44 (2012)

Miyazaki, M., Hiraki, K.: Video self-recognition in 2-year-olds. In: Proceeding of the XVth Biennial International Conference on Infant Studies. Kyoto, Japan (2006)

Mori, H., Kuniyoshi, Y.: A human fetus development simulation: self-organization of behaviors through tactile sensation. In: IEEE 9th International Conference on Development and Learning (ICDL 2010), pp. 82–97. Ann Arbor, Michigan (2010)

Murata, A., Ishida, H.: Representation of bodily self in the multimodal parietopremotor network, Chapter 6. In: Funahashi, S. (ed.) Representation and Brain. Springer, Tokyo/New York (2007)

Myowa-Yamakoshi, M., Takeshita, H.: Do human fetuses anticipate self-directed actions? A study by four-dimensional (4d) ultrasonography. Infancy **10**(3), 289–301 (2006)

Nagai, Y., Rohlfing, K.J.: Computational analysis of motionese toward scaffolding robot action learning. IEEE Trans. Auton. Ment. Dev. **1**(1), 44–54 (2009)

Nagai, Y., Hosoda, K., Morita, A., Asada, M.: A constructive model for the development of joint attention. Connect. Sci. **15**(4), 211–229 (2003)

Nagai, Y., Asada, M., Hosoda, K.: Learning for joint attention helped by functional development. Adv. Robot. **20**(10), 1165–1181 (2006)

Nagai, Y., Kawai, Y., Asada, M.: Emergence of mirror neuron system: immature vision leads to self-other correspondence. In: IEEE International Conference on Development and Learning, and Epigenetic Robotics (ICDL-EpiRob 2011), pages CD–ROM, Frankfurt, Germany (2011)

Nakano, T., Watanabe, H., Homae, F., Taga, G.: Prefrontal cortical involvement in young infants' analysis of novelty. Cereb. Cortex **19**, 455–463 (2009)

Neisser, U. (ed.): The Perceived Self: Ecological and Interpersonal Sources of Self Knowledge. Cambridge University Press, Cambridge/New York (1993)

Nishitani, N., Hari, R.: Temporal dynamics of cortical representation for action. Proc. Natl. Acad. Sci. U. S. A. **97**(0027-8424 SB – IM), 913–918 (2000)

Ogawa, K., Inui, T.: Lateralization of the posterior parietal cortex for internal monitoring of self-versus externally generated movements. J. Cogn. Neurosci. **19**, 1827–1835 (2007)

Oudeyer, P.-Y.: Phonemic coding might result from sensory-motor coupling dynamics. In: Proceedings of the 7th International Conference on Simulation of Adaptive Behavior (SAB02), pp. 406–416, Edinburgh, UK (2002)

Papousek, H., Papousek, M.: Intuitive parenting: a dialectic counterpart to the infant's precocity in integrative capacities. In: Handbook of Infant Development, pp. 669–720, Wiley, New York (1987)

Pellegrino, G., Fadiga, L., Fogassi, L., Gallese, V., Rizzolatti, G.: Understanding motor events: a neurophysiological study. Exp. Brain Res. **91**(1), 176–180 (1992)

Ray, E., Heyes, C.: Imitation in infancy: the wealth of the stimulus. Dev. Sci. **14**, 92–105 (2011)

Rizzolatti, G., Arbib, M.A.: Language within our grasp. Trends Neurosci. **21**, 188–194 (1998)

Rizzolatti, G., Sinigaglia, C., Anderson, F.: trans. Mirrors in the Brain How Our Minds Share Actions and Emotions. Oxford University Press. Oxford, New York (2008)

Russell, J.A.: A circumplex model of affect. J. Pers. Soc. Psychol. **39**, 1161–1178 (1980)

Sanefuji, W., Ohgami, H.: Responses to "like-me" characteristics in toddlers with/without autism: self, like-self, and others. In: Abstract Volume of XVIth International Conference on Infant Studies, pp. 245–246. Vancouver, Canada (2009)

Shimada, S.: Brain mechanism that discriminates self and others (in Japanese). In: Hiraki, K., Hasegawa, T. (eds.) Social Brains: Brain that Cognizes Self and Others, pp. 59–78. University of Tokyo Press, Tokyo, Japan (2009)

Takahashi, Y., Tamura, Y., Asada, M., Negrello, M.: Emulation and behavior understanding through shared values. Robot. Auton. Syst. **58**(7), 855–865 (2010)

Umilta, M.A., Kohler, E., Gallese, V., Fogassi, L., Fadiga, L., Keysers, C., Rizzolatti, G.: I know what you are doing: a neurophysiological study. Neuron **31**(1), 155–165 (2001)

Watanabe, A., Ogino, M., Asada, M.: Mapping facial expression to internal states based on intuitive parenting. J. Rob. Mechatron. **19**(3), 315–323 (2007)

Watanabe, H., Homae, F., Nakano, T., Taga, G.: Functional activation in diverse regions of the developing brain of human infants. Neuroimage **43**, 346–357 (2008)

Yamaguchi, Y.: The brain computation based on synchronization of nonlinear oscillations: on theta rhythms in rat hippocampus and human scalp EEG. In: Marinaro, M., Scarpetta, S., Yamaguchi, Y. (eds.) Dynamic Brain – From Neural Spikes to Behaviors. LNCS, vol. 5286, pp. 1–12. Springer, Heidelberg (2008)

Yoshikawa, Y., Asada, M., Hosoda, K., Koga, J.: A constructivist approach to infants' vowel acquisition through mother-infant interaction. Connect. Sci. **15**(4), 245–258 (2003)

Chapter 5
Attention and Preference of Humans and Robots

Yuichiro Yoshikawa

Abstract Receipt of response from a robot is a well-known effective factor for the formation of human preference toward the robot and has been utilized for improving human-robot interaction. In this chapter, three hypothetical phenomena regarding this preference formation are derived from recent findings of human cognitive processes and past studies: receipt of *proxy response*, *provoked attention*, and *mirrored attention*. Three psychological studies using human-robot interaction are reviewed, and thereby the ideas, effects, and possibilities of utilizing these phenomena are illustrated.

Keywords Social robot • Nonverbal channel • Attentive behavior • The feeling of being attended • Receipt of direct response • Receipt of proxy response • Provoked attention • Mirrored attention • Preference formation • Bystander robot • James and Lange • Gaze cascade effect • Heider's balance theory

5.1 Introduction

Recent advances in robotics have led to the consideration of social robots as participating in human society. Humanoid robots are expected to function as an intuitive interface between artificial system and human or to support our daily activities in social situations in virtue of their humanlike features (Kanda et al. 2004; Ishiguro and Minato 2005; Robins et al. 2004).

We have to learn the mechanisms of human sociality in order to advance robotics further in this direction, since they are the target we aim to control. The previous studies in the field of human social behavior or cognition provide us with a great deal of insights into designing social robots. The previous studies, however, are not directly used for constructing a dynamic robotic system, because they have difficulties with experimentally dealing with dynamic communication. In the conventional psychological experiment, researchers use one or a set of human subjects as stimuli for communication and manipulate them as they like. It is

Y. Yoshikawa (✉)
Graduate School of Engineering Science, Osaka University, Toyonaka, Osaka, Japan
e-mail: yoshikawa@irl.sys.es.osaka-u.ac.jp

M. Kasaki et al. (eds.), *Cognitive Neuroscience Robotics A*,
DOI 10.1007/978-4-431-54595-8_5

not, however, generally easy for researchers to do just what they aim at in the experiment, since human subjects cannot inhibit their unconscious responses. In other words, researchers cannot control human's unconsciousness, even though it may affect communication and social recognition. On the other hand, it is worth noticing that humanoid robots can be used as *controllable humans* in the experiment concerning communication. Such robots can operate in such a way that researchers can systematically manipulate the situation of interaction. Therefore, engineering studies for constructing social robots which are capable of human communication will promote scientific studies of the mechanisms of human society and vice versa (Ishiguro 2007).

Shared attention is one of the most important skills for the social development of children (Moore and Dunham 1995), and therefore robots must have it to socially interact with humans (Breazeal 2003; Imai et al. 2003). However, humans barely feel attended to by robots as they are now; the feeling of being attended seems to be the most basic component of shared attention. In order to develop more effective social robots, then it is more promising to utilize nonverbal channels to represent their attention to humans than to use explicit channels, such as verbal response (e.g., telling "I am attending to you"). This is because nonverbal channels are often used by humans, and robots may not have the ability to speak (even if they do, they may not have the opportunity to speak).

Previous studies have shown that a robot using nonverbal channels to respond to a human is effective for making him or her feel that the robot directs attention to him or her. It has been suggested that the nodding of a responsive robot leads the speaker to feel that the robot listens to him or her and it engages in conversation with him or her (Watanabe et al. 2003; Sidner et al. 2006). The responsive gaze shift of a robot gives participants a strong feeling of being looked at by it, no matter how the gaze shift is produced (Yoshikawa et al. 2006). Even the responsive use of a subtle nonverbal channel, namely, eye blinking, enhances the feeling of being looked at (Yoshikawa et al. 2007). Therefore, nonverbal responsive action is expected to be used for robots to produce attentive behavior that can be effectively recognized by humans. It is also worth noting that the experience of being directly responded to, or, more simply, the receipt of direct response, is of high potential importance for social robot study: humans, if having received a response from a robot, may interpret the robot's action in favorable ways and establish a preferred relationship with it. Since the dynamics of attention is complicated and allows for other interpretations, we should examine ways in which one can interpret other's attention.

A recent experiment in psychology shows that attentive behavior plays a crucial role in preference formation (Shimojo et al. 2003). In this experiment, subjects are presented with two pictures on the screen and asked to choose a preferable one. Their focal points are measured during their choosing task. Attentive behavior (i.e., looking at either side of a picture) precedes choice making. In other words, subjects prefer one picture to the other because they have attended to it. We take this to suggest a different potential way to make humans prefer robots: by biasing human's attentive behavior toward the target of social cognition, or more simply, by provoking human's attention to the target.

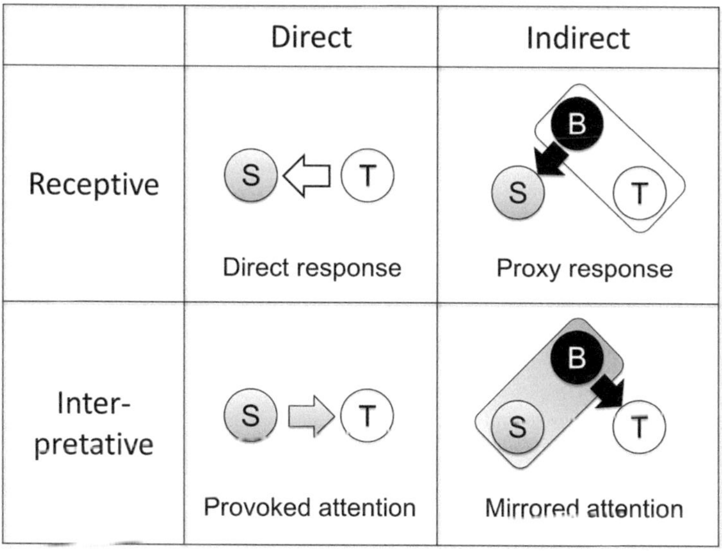

Fig. 5.1 Types of subjective interpretation of attention in social situation: S, T, and B indicate the subject of cognition, the target of cognition, and the bystander, respectively

The two ways of preference formation toward robots so far mentioned, the receipt of direct response and provocation of attention, are based on attentive behavior. A different way is to utilize social influence. It is the central assumption of social psychology that a person's cognition is easily influenced by other person's cognition (Heider 1958). Therefore, the process of preference formation can be mediated by a third agent. If we can build a humanoid robot with nonverbal capacities that can participate in communication with humans in a less obtrusive way, it may be able to mediate their preference.

The ways of preference formation toward robots are modeled on preference formation toward persons. An important question to be addressed is how humans form preferences toward others on the basis of attentive behavior. Figure 5.1 illustrates the basic categorization of possible ways to interpret social relationship in dyad and triad situations. In this chapter, three ways are examined in view of experimental results involving human-robot interaction: provoked attention, proxy response, and mirrored attention. Preference affects attention, and attention affects preference. As a first step toward the explication of the interplay between attention and preference, in Sect. 5.2, we review a study on human-robot interaction and see how a similar chicken egg structure of behavioral and cognitive processes appears in social situations (Yoshikawa et al. 2008). This study is concerned with the issue of provoked attention and more specifically with the questions of how the subject interprets his or her own occasional responsiveness toward a robot via nonverbal channels and how the subject's interpretation biases his or her preference formation toward the robot. By examining this study, we reveal to what extent it is feasible to make humans prefer robots. As it will turn out, social situations necessarily restrict human's unconscious responses to robots.

By extending the two direct ways mentioned above, we can sophisticate and construct two social ways of preference formation toward robots: we call them "proxy response" and "mirrored attention." Proxy response is the phenomenon in which the subject mistakes the responsive attention of a third agent for that of the target agent. Mirrored attention is the phenomenon in which the subject cognizes the attention of a third agent toward the target agent as the subject's own cognition of the target. In order to construct these ways, first, we should examine how or whether a third agent can use nonverbal channels to influence the two direct ways of preference formation toward a robot (or a human). We review two studies that deal with this question. In Sect. 5.3, then, we describe the examination of nonverbal response by a bystander robot, which can be seen as a social implementation of the effect of the receipt of direct response. In Sect. 5.4, we describe the examination of the effect of observing nonverbal interaction between a robot and a person, which can be seen as a social implementation of the effect of provoked attention to the target robot (Shimada et al. 2011). The phenomena examined here are expected to hint at how to theorize collective behavior of social group robots for the purposes of mediating humans and robots and understanding the dynamics of human sociality along the dimension of attentive behavior and preference formation.

5.2 Provoked Attention

As have been argued in psychology after the James-Lange hypothesis, one's feelings might originate from one's physiological responses (James 1884). For example, it might be more appropriate to interpret people as feeling sad because they are crying than to interpret them as crying because they are sad. A similar phenomenon is called the "gaze cascade effect," in which an inverse causal relationship obtains between feeling and action or, more properly, between preference formation and attentive behavior (Shimojo et al. 2003). For example, it might be more appropriate to interpret someone's preference for an object because he or she is looking at it than to believe that he or she is looking at it because he or she prefers it.

The question of interest is how or whether this kind of inverse causal relationship appears in preference formation for a communication partner in a social situation. In the forward-looking (normal) understanding of causality, the preference for a communication partner is the cause, and the attentive behavior to the communication partner is the effect. However, this understanding is not always correct, and the inverted causal relationship might obtain: the preference for a communication partner might be the cause of the occasional behavior of responding to him or her. This implies the possibility of utilizing a subject's attentive behavior to other persons or robots in order to make the subject prefer them. To test this hypothesis, we have built a robot system and used it to manipulate the subjects; when they

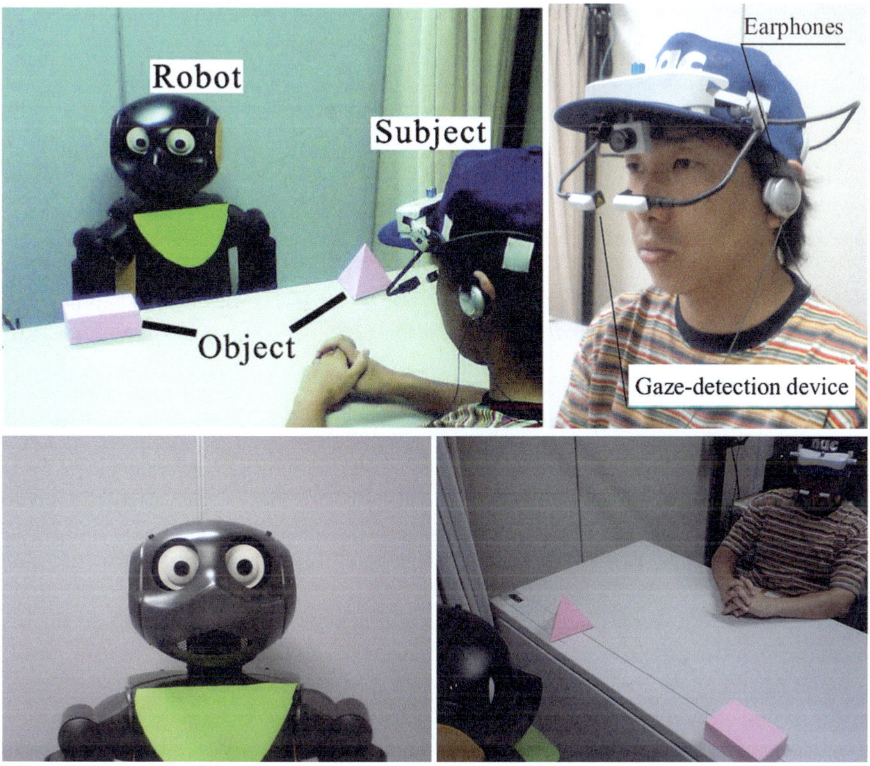

Fig. 5.2 Experimental situation for provoked attention

respond to the robot, they are led to experience a gaze interaction without any voluntary intention for interaction. In this section, we describe this experiment in detail and see how or whether the subjects regard a robot as more communicative than otherwise when they have experience of responding to it.

5.2.1 Method

1. *Subjects*: 66 naïve Japanese participants (age: M = 23.8, SD = 2.8). Participants were assigned to one of the three conditions. We obtained data from 11 male and 11 female participants under each condition.
2. *Apparatus*: A humanoid robot (Robovie-R2, ATR Robotics) was used for the experiment (see Fig. 5.2). Although it had 17 DOFs in total, only six DOFs were used, i.e., only the pan and tilt axes of its neck and the pan and tilt axes of both of its eyes. Two paper blocks with plain surfaces were placed on a table between the robot and the participant.

A gaze detection device (EMR-8B, Nac Image Technology, Inc.) was used to measure the subject's foci of attention. A host computer received gaze data: a sequence of focal points of the participant's eyes, from the gaze detection device, and images from a camera mounted on the participant's head. By using these data, the computer judged whether the participant was looking at the robot or at the two other objects. The computer also sent sound signals to the earphone the participant was wearing, to inform him or her of the timing of the relevant action.

3. *Procedure*: An experimenter led a participant to the room and told him or her to sit on a chair facing a table across which a robot was standing but hidden by a curtain. The participant was told that he or she would attend three 1-min sessions where he or she had to look at the objects on the table when he or she heard sound signals from the earphones or look at the robot when he or she did not. The patient was also told that he or she had to answer the questionnaire on feelings toward the robot after each 1-min session.

   Before interacting with the robot, the experimenter asked the participant to wear the gaze detection device and calibrate it. The experimenter did not explain that the device may be used to control the gaze of the robot, but merely that it was used to analyze human-robot interaction. The participant then practiced changing gaze in response to sound signals. Note that the curtain was opened during the practice so that the participant could habituate to the robot. The robot did not move at all during the practice.

   After the participant had understood what he or she had to do in the sessions, the experimenter left the room and the first session was started. Just before the 1-min gaze interaction had finished, the experimenter silently returned to the room entrance located behind the participant. The robot suddenly directed gaze to the experimenter at 54 s after the beginning of the session. This was for finding out whether the robot was able to draw the participant's gaze. When the participant looked back toward the experimenter or 6 s passed from the robot's attempt at gaze drawing, the experimenter asked the participant to answer the questionnaire about the impressions he or she had formed through the session. Following a 2-min interval after the completion of the questionnaire, the second session was started in the same way as in the first one and was followed by the third session in the same way.

4. *Measurement*: Images were obtained from the frontal camera on the cap the participants wore, and gaze data were collected by using the gaze detection device, in order to calculate which objects they were looking at during the sessions and the gaze-drawing tests at the end of the sessions.

   The questionnaire consisted of 15 questions about the feelings the participants had formed through the sessions. The first seven questions were given positive or negative answers. The other eight questions contained conjugate adjective pairs concerning the characteristics of the robot. For both types of questions, they graded on a five-point scale the degrees to which their answers matched their feelings; a high value corresponded to a strong feeling about the right-side adjective and a low value to a strong feeling about the left-side adjective. Note that the same questionnaire was used for all the sessions.

5. *Stimuli*: The gaze interaction between the participant and the robot was not only controlled by the robot gaze but also by it signaling the participant to change gaze. The participant was instructed to look at either one of the objects on the table when he or she heard a sound from the earphones but otherwise to look at the robot's face. Therefore, the timing to change the participant's gaze could be roughly controlled with the signal sound. The signal sound was emitted every 12 s and lasted for 6 s in each session. In other words, the participants were controlled to look alternately at the robot's face and either one of the objects for every 6 s.

   The robot looked alternately at the participant's face and one of the two objects. It randomly selected and looked at either one of the objects. The way it looked at a target was manually implemented by adjusting the posture of its neck and eyes. The timing to switch the target was controlled differently in three conditions:

   - *Responding condition*: The robot's gaze is controlled so that it shifts its gaze just before the main computer sends the signal sound and the participant shifts his or her gaze. The robot's gaze shift precedes the participant's by 0.5 s. The participant is expected to have experience as of responding to the gaze movement of the robot.
   - *Being-responded condition*: The gaze of the robot is controlled so that it shifts its gaze 0.5 s after the participant shifts his or her gazes. The onset timing of the participant's gaze shift is detected by the gaze detection device. We use the same parameter of latency as we used in the previous work on the effect of being responded to by a robot (Yoshikawa et al. 2006). The participant is expected to have experiences as of being responded to by the robot.
   - *Independent condition*: The gaze of the robot is controlled so that it shifts its gaze independently of the participant's gaze movement. The robot's gaze shift happens 3.0 s after the auditory signal is emitted and terminated. The participant is expected to have no experience as of either responding to or being responded by the changes in the robot's gaze.

6. *Predictions*: If participants had experience as of responding to the robot, they would regard it more strongly as a communicative being. We thus predicted that participants under the responding condition would evaluate higher the answers about the relevant topics to a communicative being, such as their attributing intentions or agency to the robot and its friendliness, compared with participants under the other two conditions. Furthermore, one's responding behavior strengthened one's tendency to exhibit responsive behavior to the robot. We also predicted that the robot would draw the gaze of the participants under the responding condition more frequently in the gaze-drawing test at the end of each session.

5.2.2 Results

1. *Manipulation check*: To check to what extent gaze interaction could be controlled in the predicted way, we calculated the average response latency of the robot under all the conditions using valid gaze data. Under the being-responded and independent conditions, the average latency of the robot's response to the participant's gaze shift had been calculated. For the responding condition in which the participant was assumed to have experience as of responding to the robot, we calculated the average latency of the participant's response to the robot's gaze shift. The calculated values showed that the manipulation of response latency was successful (responding ($M = -0.55$, $SD = 0.26$), being responded ($M = 0.52$, $SD = 0.12$), independent ($M = 3.06$, $SD = 0.63$)).
2. *Subjective effects*: Participants answered 15 questions about each session. We first compared the average scores on the first eight questions in the three conditions. These questions were designed to evaluate the participants' feelings toward the robot. Using ANOVA with data from the first session for all participants, we found a weak statistical tendency to answer differently the first question "Did you feel as if the robot had its own intentions?" ($F(2, 60) = 2.67$ and $p < 0.1$) and the second question "Did you feel as if the robot were an animate entity?" ($F(2, 60) = 2.73$ and $p < 0.1$). Note that data from three participants in the being-responded condition were excluded from the analysis, because they contained more than two cases of 1.0-s or longer response latencies of the robot gaze. Post hoc analysis indicated marginal differences: the scores for the first question in the responding condition were more positive than those in the independent condition (Hochberg: $p < 0.1$), and the scores for the second question in the being-responded condition were more positive than those in the independent condition (Tamhane: $p < 0.1$).

   Principal factor analysis with a varimax rotation was conducted on the scores on the eight conjugate adjective pairs for the characteristics of the robot by using data in the first sessions. Note that data from three participants in the being-responded condition were excluded from the analysis, because they contained more than two cases of unexpected latencies of the robot gaze. We found three factors that represented 59.6 % of variances in the data. Since the first factor mainly concerned the scores on such pairs, as trustworthy vs. anxious (0.68), considerate vs. inconsiderate (0.76), and cold vs. warm (−0.83), we called it "friendliness" (values in brackets indicate factor loadings). We called the second factor "self-reliance" since its main gradients were worrywart vs. dispassionate (0.66), unconfident vs. confident (0.60), and expert vs. amateur (−0.54); we called the third factor "thoughtfulness" since its main ingredient was just the pair of thoughtful vs. impulsive (0.70). By ANOVA ($F(2, 60) = 3.43$ and $p < 0.05$) and a post hoc test using Hochberg's (GT2) method, we found a significant difference in scores on the principal factors of robot characteristics: the feeling of friendliness in the responding condition was stronger than in the independent condition ($p < 0.05$). We did not find any tendencies for the scores on self-reliance and thoughtfulness.

3. *Behavioral measure*: The robot shifted its gaze toward the room entrance to draw the participant's gaze at the end of each session, and we obtained results from 198 trials of the robot's gaze drawing. We regarded the robot to succeed at gaze drawing if the participant's gaze was directed toward the entrance side within 5 s after the onset of the robot's gaze-drawing behavior. The ratio of the success of the robot's gaze drawing was 33 % in the responding condition, 15 % in the being-responded condition, and 21 % in the independent condition. Note that 21, 18, and 9 cases were respectively excluded from the analysis of these conditions, since, in these cases, the gaze detection device had failed to detect more than 30 % of 5 s or the condition for the success of the robot's gaze response had not been met in the being-responded condition. Although, as we expected, the ratio of gaze drawing was highest under the responding condition, the tendency was not significant ($\chi^2 = 4.41$, df $= 2$, $p = 0.11$). This might be because many participants assumed that the session was not over at the time of the gaze-drawing test, and they therefore inhibited their gaze movements; there was no signal sound in the gaze-drawing test.

5.2.3 Conclusion

Through the experiments, we found marginally significant results that indicate that the participant's feelings about the communicativeness of the robot can be positively strengthened by their experiences as of responding to it. But we had not yet found significant effects on the participant's spontaneous response to the robot. Therefore, the results marginally support the hypothesis that a feeling possibly leading to preference formation can be caused by occasional experiences of responsive-attentive behaviors. Further experiments should be conducted by releasing experimental constraints on the subject's action, in order to check whether the feeling enhances the subject's spontaneous response. The method for inducing person's attentive behavior to other persons or robots and controlling his or her preference is important, as this and further experiments will show. To fully realize its potential, a method for controlling human behavior in social situation should be developed in the future research.

5.3 Proxy Response

Heider's balance theory is important for considering interpersonal cognition of more than two persons (Heider 1958). We will theorize the phenomenon of what we call the "nodding confusion" on the basis of Heider's idea that cognitions of related agents influence each other. Heider represents relationships between two persons in a triad as either positive or negative and triadic relationships as eight possible combinations of relationships between two persons. A relationship among persons is regarded as positive not only if one of them has or at least observes a positive sentimental attitude but also if they form a unit relationship where they are treated as belonging to one class. Heider classifies how a subject feels about a triad into two states: the balanced state in which the subject feels comfortable and the unbalanced state in which the subject feels uncomfortable. The core of balance theory is that a subject is biased to form or evaluate preferences so as to balance the resultant triadic relationship.

What we call the "nodding confusion" can be understood by considering the three-person interaction where nodding is used as an attentive behavior. Let S, L, and B be a speaker, a listener, and a bystander, respectively, and suppose that B is watching a conversation between S and L. If B nods in response to S, then the previous studies on the receipt of response (Watanabe et al. 2003; Sidner et al. 2006) predict S is biased to believe that his or her words are accepted by B, and so the relationship between S and B is positive. L and B are physically close to each other and appear to have the same role of a listener. Since they look close, S is considered to regard the relationship between L and B as positive, in balance theory. In a situation like this, balance theory predicts that if S's cognition of the relationship between S and L is biased to be positive, S feels comfortable with the triadic relationship. Can S be self-convinced of this biased cognition? Although L's nodding might be regarded as the cause of the biased cognition of a relationship, this is not necessarily true. Here, we conjecture that since cognition of nonverbal behavior, such as other's nodding, is highly implicit, cognition of L's nodding can be reorganized as if it has been observed. We call this illusion phenomenon the "nodding confusion" (see Fig. 5.3).

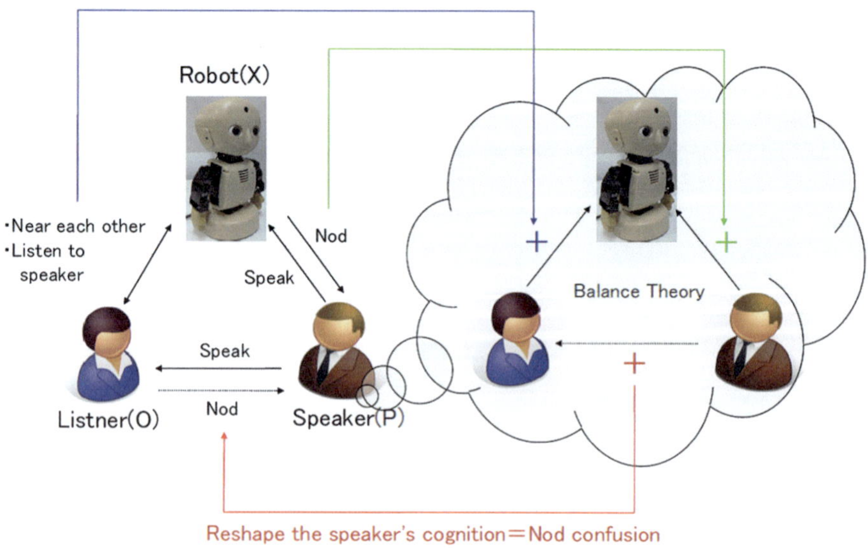

Fig. 5.3 Nodding confusion

If the hypothesis of the nodding confusion is true, we can bias cognition of interpersonal relationship merely by using a robot that nods to the speaker's talk. It is predicted that even if L (interlocutor) does not nod at all, S (subject) can be deceived into believing that he or she has seen L's nodding. For such self-deception or reorganization makes cognition of triadic relationship comfortable in the light of Heider's theory. That is, it may be possible to utilize a bystander robot to show certain attentive behaviors and elicit responses in the way that promotes preference formation for the target agent. In this section, we describe two new experimental results pertaining to the hypothesis of the nodding confusion. In Sect. 5.3.1, we discuss an experiment of human-robot-robot interaction to see how or whether proxy response can promote preference formation for a robot. In Sect. 5.3.2, we discuss an experiment of human-human-robot interaction to explore the possibility that a bystander robot can socially apply proxy response.

5.3.1 *Are Robot's Nods Confused with Another Robot's?*

To examine whether the nodding confusion is caused by two robots, we have conducted an experiment of human-robot-robot interaction. In this experiment, a human subject is asked to talk to two robots, and only one of them may nod in response to the subject's words. The question is whether the subject confuses the nodding responses of one robot with those of the other robot.

5.3.1.1 Method

1. *Subject*: 24 Japanese adults (12 males and 12 females), whose ages ranged from 18 to 23. We adopted a between-subjects design. Twelve subjects (five males and seven females ($M = 19.9$, $SD = 1.6$ [years])) were assigned to the B-Nod condition where a target robot did not nod at all during the experiment and a bystander robot nodded in response to the subject. Another 12 subjects (six males and six females ($M = 20.4$, $SD = 1.2$ [years])) were assigned to the B-unNod condition where both the target and bystander robots did not nod at all during the experiment.
2. *Apparatus*: Two robots called "M3-Synchy" and a picture of curry rice were placed on a table (see Fig. 5.4). M3-Synchy was an upper-body humanoid robot of about 30 cm in height, and it was designed to be so small and anthropomorphic that it could be easily introduced into communication among humans without disturbing them. It could use its eyes and head to exhibit its attention or nodding gesture. The distance of the two robots was 35 cm, and the distance between a subject and each robot was 45 cm. The subject could see both robots at the same time. We called the left-side and right-side robots the "target" robot (T-robot) and the "bystander" robot (B-robot), respectively, and examined the effect of the B-robot on the subject's cognition about the T-robot. The subject wore a headset

Fig. 5.4 Experimental scene for nodding confusion between robots

to hear the robots' utterances. Two cameras were placed in a 3 by 2.5 meter experimental room to record the conversation.

3. *Procedure*: Each subject was asked to explain to the robots how to cook curry rice. The subjects were told that the purpose of the experiment was to examine how humans feel when they talked to robots. Before the experiment, the subjects had a chance to read a recipe for curry rice and practice explaining it toward a wall. Then, the experimenter brought the two robots on the table and left the room, and then the subject started the explanation. Once the explanation was done, the experimenter brought out the robots and asked the subject to answer the questionnaire about his or her feeling.
4. *Stimulus*: Two robots, the T-robot (target robot) and the B-robot (bystander robot), were used as listeners of the subject's explanation. The B-robot behaved differently, depending on the conditions: B-Nod or B-unNod. On the other hand, the T-robot behaved in the same way in both conditions.

   - The *T-robot* was placed on the left side of the subject. It did not nod during the subject's explanation. Instead, it showed gaze behavior and looked as if it looked alternately at the picture of curry rice for 3 s and at the subject's face for 7 s.
   - The *B-robot* was placed on the right side of the subject. In the B-Nod condition, it showed nodding behavior when it detected a short pause in the subject's voice. Note that the B-robot nodded only once in three times when it detected a short pause, to avoid the impression that it nodded too frequently. By contrast, it did not nod in the B-unNod condition. It showed gaze behavior in the same way as the T-robot did in both conditions. It chose one of the two types of nodding movements at random: slowly nodding once or quickly nodding twice.

5. *Measurement*: We evaluated how the subject felt about each robot by using a questionnaire with a seven-point Likert scale. We asked the subjects to answer the following questions about each robot. The label X in the questions was substituted by either the "B-robot" or "T-robot," and there were two sets of questions.

   Q1-X Did the robot X nod to your words? (nodding)
   Q2-X Did the robot X listen to you? (listening)
   Q3-X Did the robot X understand your explanation? (understanding)
   Q4-X Did the robot X feel close to you? (robot to human closeness)
   Q5-X Did you feel close to the robot X? (human to robot closeness)

   To exclude cases where the subjects completely failed to explain, we asked them to evaluate their performance with the question (Q-S): Did you succeed in the explanation?
6. *Predictions*: The relationship between the two robots was regarded as positive in Heider's balance theory, because they looked to have a unit (very close) relationship to one another, due to their physical closeness and similarities of appearance. Therefore, if the hypothesis of the nodding confusion were true, the subject confused one robot's nodding with the other robot's, and accordingly the score for Q1-T in the B-Nod condition was higher than in the B-unNod condition. In addition, several aspects of the subject's cognition were subject to the consequent effects that were generally considered to be induced by the feeling of being nodded by the interlocutor in human-human communication. Q2-T, Q3-T, Q4-T, and Q5-T were relevant to those effects, and we predicted that the scores for them would be higher in the B-Nod condition than in the B-unNod condition. However, nodding by the bystander robot may not be strong enough to sufficiently affect some aspects of the subject's cognition. For such aspects, the consequent effects would not be observed.

5.3.1.2 Result

1. *Manipulation check*: In the B-Nod condition, the average number of robot's nodding was 7.9 ($SD = 2.1$). Since all the subjects scored more than 1 for Q-S, the question of whether they were able to explain well, we included all data in the analysis.
2. *Subjective effects*: The medians of the scores for the B-robot and T-robot are shown in Table 5.1 with p-values. We could see that the medians of all the scores in the B-Nod condition were higher than in the B-unNod condition. The results of the Kolmogorov-Smirnov test suggested that the score for any question in either condition was not normally distributed, with an exception of Q5-T. Therefore, we used a parametric test for Q5-T and Kruskal-Wallis nonparametric tests for other questions, to compare scores. We selected the same Kruskal-Wallis statistics as we do in the next section.

First, we checked the scores for the B-robot (the bystander robot). We found that the B-robot was more strongly felt to nod in the B-Nod condition than in the B-unNod condition (see Q1-B "nodding"). This indicated that the manipulation of

Table 5.1 Medians and p-values of questionnaire scores in the experiment of the nodding confusion between robots

	Question	B-unNod	B-Nod	P-value
B-robot	Nodding	4.5	6	$P<0.01$
	Listening	4.5	6	$P<0.01$
	Understanding	4	5	*n.s.*
	R to H closeness	4	6	$P<0.01$
	H to R closeness	4	5	$P<0.1$
T-robot	Nodding	4	5	$P<0.05$
	Listening	3	5	$P>0.05$
	Understanding	3	3.5	*n.s.*
	R to H closeness	2.5	3.5	$P<0.05$
	H to R closeness	3.5	5	*n.s.*

the robot's nodding was successful. In accordance with this result, the B-robot was felt strongly to listen to the subject's explanation when it nodded to the subject (see Q2-B "listening"). These results indicated that the nodding of the robot was sufficiently functioning in this experiment (similar results were obtained in different experiments, and this should not be surprising (Watanabe et al. 2003)). However, the nodding of the robot was not strong enough to make the subjects feel as if it understood their explanation (see Q3-B "understanding"), although the significance level was not very low but marginal. The nodding manipulation was strong enough to make the subjects feel to have a positive relationship with the B-robot. The feeling was passive, i.e., they felt as if the robot was close to them (see Q4-B "R to H closeness"), and also active, i.e., they felt as if they were close to the robot (see Q5-B "H to R closeness").

Second, we checked the scores for the T-robot (the target robot). We found that the T-robot was more strongly felt to nod (see Q1-T "nodding"), to listen (see Q2-T "listening"), and to be close (see Q4-T "R to H closeness") in the B-Nod condition than in the B-unNod condition. Furthermore, the subjects in the B-Nod condition strongly felt close to the T-robot (see Q5-T "H to R closeness"). Because the T-robot did not show any nodding behavior in this experiment, this effect of the increase in feeling came from the B-robot, and hence it was an instance of the phenomenon of the nodding confusion. On the other hand, there was no significant difference in the felt understanding by the bystander robot between the B-Nod and B-unNod conditions (see Q3-T "understanding"). As was predicted in the hypothesis section, this may be because the nodding function of the B-robot was weak at affecting the aspect of understanding (see Q3-B "understanding").

5.3.2 *Are the Robot's Nods Confused with a Next Person's?*

To examine whether the nodding confusion appeared between a human and a robot, the subjects were asked to talk to a human interlocutor with and without holding a nodding robot on their laps (see Fig. 5.5). The interlocutor was an experimental confederate who was trained not to nod or smile while the subjects were talking.

Fig. 5.5 Experimental scene of the nodding confusion between a human and a robot

5.3.2.1 Method

1. *Subject*: 36 Japanese adults (17 males and 19 females), whose ages ranged from 18 to 29. Before the experiment started, they were exposed to and talked with the robot used in the experiment for the practice purpose. We adopted a between-subjects design. Twelve subjects (five males and seven females (M = 23.3, SD = 2.7 [years])) were assigned to the Nod condition where nodding robots were used for the experimental conversation; 12 subjects (six males and six females (M = 22.5, SD = 1.8 [years])) were assigned to the Without condition where robots were not used; and the remaining 12 subjects (M = 20.9, SD = 1.4 [years])) were assigned to the unNod condition where robots were used but did not nod.
2. *Apparatus*: Two cameras were placed in a 3 by 2.5 m experimental room, and the conversation was recorded by them. The distance between the subject and the interlocutor was about 1 m, and there was a small round table between them. Two sets of five topic cards that described the questions the interlocutor would ask were placed on the table. The subject and the interlocutor wore headsets to hear their utterances.
3. *Procedure*: Each subject had a conversation with an interlocutor about the topic selected by a topic card; the interlocutor and the subject each had a set of topic cards and asked and answered a question to each other. Each subject was given a list of items to be asked about (topics) and had 5 min to prepare for answers, prior to the experiment. Then, the interlocutor entered the room and gave the subject the robot that was used in the Nod condition. All the topic cards were placed facedown at the beginning of conversation. The subject turned up the first card, asked the question on it, and listened to the interlocutor's answer. The interlocutor asked the first question on his card and listened to

the subject's answer. After this cycle was repeated five times, the subject answered a questionnaire about the feelings he or she formed in and through the conversation.

4. *Stimulus*: The subjects were asked to communicate with an experimenter about prepared topics. A 24-year-old Japanese male was assigned the role of an interlocutor for the subjects. He had the same questions from all the subjects and answered to them in the same ways. Note that he was trained not to nod or smile in response to the subjects. The topic cards specified precisely what questions were to be asked. We selected ten easy questions, for example, "please tell me what recently surprised you" or "please tell me three things you did this summer." The questions and their order were the same among the subjects.

   There were three conditions, Nod, Without, and unNod conditions, where the robots behaved differently:

   - *Nod condition*: The subject and the interlocutor sit on different sides of the room and talk to each other with a robot on their laps. The robot on the subject's laps looks at the target agent, either the person or the company robot on the other side. The target agent for each robot is switched every 10 s. The robots are programmed to nod in response to the subject's words on the other side. If a breath group is detected by the microphone each subject wears and more than 3.0 s have passed from the last nod, one of the two types of nodding is selected randomly: slowly nodding once or quickly nodding twice.
   - *unNod condition*: The subject and the interlocutor sit on different sides of the room and talk to each other with a robot on their laps. The behaviors of the robots are the same except that they do not nod.
   - *Without condition*: The subject and the interlocutor sit across from and talks to each other, using topic cards.

5. *Measurement*: We evaluated the conversations by analyzing the questionnaire answers and video records of the subjects and the interlocutor. In the questionnaire, the subjects graded on a 7 discrete scale the degrees to which they positively answered the questions concerning the feelings they had formed in the conversation of the experiment. There were two types of questions: questions about the feelings the subjects had when they were talking and questions about the general feelings they had formed through the conversation. Note that 1 means "strongly positive" and 7 means "strongly negative." Below is a full list of questions:

   - Questions about the feelings during the subject's talk:

     Q1-T Do you think the interlocutor nodded in response to you? (nodding)
     Q2-T Do you think the interlocutor listened to you? (listening)
     Q3-T Do you think the interlocutor understood you? (understanding)
     Q4-T Did you answer the interlocutor's questions well? (talking)

- Questions about the general feelings formed through the conversation:

  Q1-A Do you think that the interlocutor felt close to you? (E to S closeness)
  Q2-A Do you feel close to the interlocutor? (S to E closeness)
  Q3-A Were you able to relax during the conversation? (relax)
  Q4-A Do you think that you could be a friend of the interlocutor? (real influence)

6. *Prediction*: We supposed that the relationship between a person and a robot on that person's laps was positive according to Heider's balance theory, because they seemed to have a unit relationship, due to their physical closeness. Then, if the hypothesis of the nodding confusion were true, the nodding of the robot might be confused with the person's. Accordingly, the score for Q1-T might be higher in the Nod condition than in the unNod and Without conditions. As with the previous experiment, the consequent effects on the aspects of the subject's cognition of the interlocutor were evaluated in terms of the increases in scores for listening (Q2-T) and understanding (Q3-T). In accordance with the increases in these scores, other scores (Q1-A, Q2-A, Q3-A, Q4-A) pertaining to the feelings about the entire conversation and the relationship between the subject and the interlocutor would be more positive. In addition, since these positive changes appeared also in the behavioral measure of how the subjects became active in their speech, the subject would have a longer conversation in the unNod condition than otherwise.

5.3.2.2 Result

We used the Tukey-Welch test to compare scores in the three conditions. We used F-statistics for the cases where data were normally distributed and Kruskal-Wallis statistics for the cases where they were not.

1. *Manipulation check*: By analyzing the video records, we confirmed that the interlocutor neither nodded nor smiled when the subjects were talking. We also checked whether the interlocutor talked in the same way in different conditions. There were no significant differences in the time he spent for his talk in the three conditions (Nod ($M = 134.4$, $SD = 9.8$ [sec]), Without ($M = 138.4$, $SD = 7.8$ [sec]), unNod ($M = 137.3$, $SD = 6.3$ [sec]); $F(2; 33) = 0.8$(n.s.)). Therefore, the interlocutor's behavior was successfully controlled in these between conditions.
2. *Subjective impression*: The medians of the scores for questions are shown in Table 5.2 with p-values. We analyzed the scores for Q1-T and thereby how the subjects felt about the interlocutor's nodding. The Tukey-Welch test for Q1-T indicated a marginal difference in the three conditions with the significance level of 0.1 ($F(2; 33) = 3.13$). The subsequent tests indicated that the scores in Nod ($F(1; 33) = 5.57$) and in unNod ($F(1; 33) = 4.80$) were larger than the scores in Without.

Table 5.2 Medians and p-values of questionnaire scores in the experiment of the nodding confusion between a human and a robot

Question	Nod	Without	UnNod	P-value
Nodding	5.5	4	4	Nod > Without ($p<0.1$), Nod > unNod ($p<0.1$)
Listening	6	5.5	6	*n.s.*
Understanding	5.5	5	5.5	*n.s.*
E to S closeness	5	4	3.5	Nod > Without ($p<0.05$), Nod > unNod ($p<0.05$)
S to E closeness	6	5	5	Nod > Without ($p<0.1$), Nod > unNod ($p<0.1$)
Relax	6.5	5	5	Nod > Without ($p<0.1$)
Befriended	5	4.5	4	*n.s.*

We also analyzed the general feelings the subjects formed through the conversation. We found a significant increase in the scores for Q1-A (E to S closeness) with the significance level of 0.05, which was intended to measure the subject's evaluation of the interlocutor's feeling of closeness to the subject ($\chi^2(2)=10.5$). The subsequent test revealed that the scores were highly significantly larger in the Nod condition than in the Without ($\chi^2(1)=6.75$) and unNod ($\chi^2(1)=8.00$) conditions. Furthermore, there were also marginal differences in the scores for Q2-A (S to E closeness) only with the significance level of 0.1 ($\chi^2(2)=5.21$) and Q3-A (relax) ($\chi^2(2)=5.42$ ($p<0.1$)). The subsequent test indicated that the subjects in the Nod condition may feel closer to the interlocutor than those in the Without ($\chi^2(1)=3.52$) and unNod ($\chi^2(1)=4.20$) conditions and that the subjects in the Nod condition were more relaxed than those in the Without ($\chi^2(1)=4.69$) condition.

5.3.2.3 Conclusion

Provided that the interlocutor succeeds in inhibiting his nodding, the tendency of the subject's stronger feeling of being nodded (Q1-T) in the Nod condition can be regarded as an instance of the nodding confusion. We can therefore conclude that the subjects confuse the nodding of the robot with that of the interlocutor's, although the significance level is not high. We also confirm that there are increases in certain aspects of general positive feelings toward the interlocutor. This can be interpreted as supporting the idea that the nodding of the robots positively promotes human communication, as opposed to the idea that they interfere in a human communication and humans take a longer time for the conversation. We have yet to analyze what difficulties the subjects face in the conversation or do further experiments with questions on how the subjects think about such difficulties.

These results indicate the possibilities of mediating human communication by introducing small humanoid robots. In order to realize the possibilities, of course, we need to address further questions concerning, e.g., the possibility of using a variety of modalities (including but not restricted to nodding) to promote human communication, the optimization of the timing of robot's response, and the application to more complicated cases with more robots and/or more humans.

5.4 Mirrored Attention

The remaining social route for preference formation is mirrored attention. It has been considered that human cognition is easily influenced by other persons. Japanese adults tend to fail to recognize the facial expression of a person if the person is surrounded by persons with different facial expressions (Masuda et al. 2008). According to Heider's balance theory, human attitudes in a social situation are biased by the relationships among the subject of cognition, the target person of cognition, and the third-party person (Heider 1958). A typical bias is to mirror the attitude of a reliable person. The relevant question here is whether preferences are sufficiently affected by mirroring others' nonverbal attentive behavior.

In this section, we report an experiment of human-human-robot interaction. For this experiment, we use a scenario of a hypothetical job interview as an example of triadic communication, where two agents talk with each other and a third agent listens to their conversation. Human participants, a subject and a "confederate" (a person who is aware of the nature of the experiment and helps to carry it out), play the roles of the main speaker and the main listener, respectively; a robot (an android called Repliee Q2) plays the role of the sub-listener. The eye contact between the confederate and the robot is controlled differently in different conditions: the confederate makes eye contact several times in the EC condition and makes no eye contact in the NEC condition. The experiment uses a post-interaction questionnaire to examine how the feelings of the subjects are affected by the eye contact between the two other agents by using a post-interaction questionnaire. Furthermore, it is worth examining whether Heider's theory can predict how a subject's observation of nonverbal interaction between others influences his or her social cognition of them.

5.4.1 Method

1. *Subjects*: 30 Japanese adults (ages: M = 21.2, SD = 2.0[year]) hired through a temporary employment agency. Condition 1 was performed with 15 subjects (eight males and seven females) and condition 2 with the remaining 15 subjects (eight males and seven females).
2. *Apparatus*: An android called "Repliee Q2" was used in this experiment (Fig. 5.6). It had a very humanlike appearance and resembled an actual Japanese woman. Its actions were programmed in advance to mimic those of a store clerk, including bowing and looking at a speaker (i.e., a customer at the store). The actions were triggered at an appropriate timing by the operator in a remote room who monitored interaction in the experimental room via two cameras and a microphone installed therein. The android (seated on a stool), a round table, and two chairs were placed in the experimental room (3×3.7 [m]).
3. *Procedure*: All instructions for the experiment were given to each subject by an instructor before he or she entered the experimental room where the android was

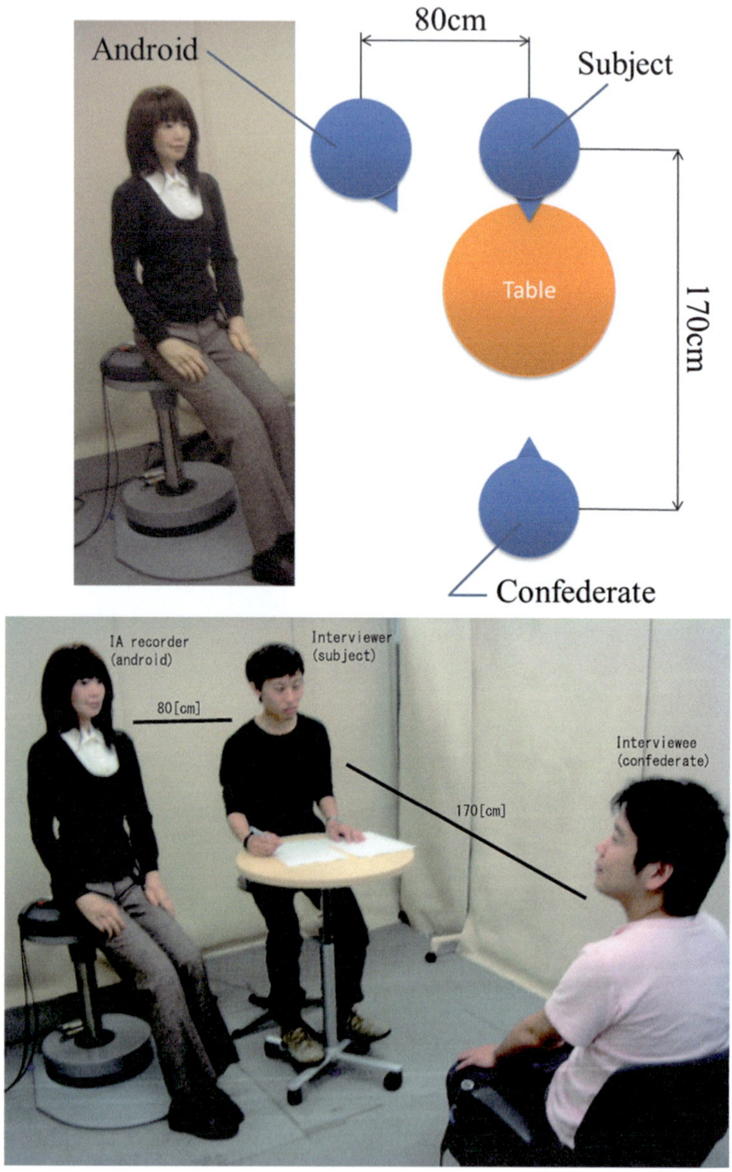

Fig. 5.6 Experimental scene for the influence of observing others' eye contact

waiting. After moving to the room, the instructor asked the subject to sit on a chair next to the android. After the instructor left, the confederate entered the room and sat on a chair in front of the subject. Then, the subject started to ask questions that were listed in a document placed on the table. The subject listened to the confederate's answers and evaluated them on a seven-point scale (this was a dummy task). There were eight questions which typically appeared in a real job interview. For example, "What kinds of things have you learned so far?" or "What kind of work do you want to do?" Having asked all the questions, the subject answered a post-interaction questionnaire, and it is designed to evaluate the subject's feelings about the interaction, including the impression of the confederate and the android.

Before each subject entered the experimental room, the instructor told the subject that the purpose of the experiment was to evaluate a mannequin-like device called the "IA recorder," a next-generation integrated chip (IC) recorder furnished with a human appearance that was easily acceptable to humans. The subject was told to read questions listed in a document on the table and ask them to the confederate who played the role of an interviewee. The subject was also told to score only the confederate's answers with high scores (dummy task) on a seven-point scale. This was needed for positively biasing the attitude of the subjects toward the confederate. Thus, the subject only needed to decide whether to give 5, 6, or 7 points for each answer.

4. *Stimulus*: The confederate was trained to behave in the same way in the two conditions. In answering the questions from the subject, he sometimes turned his gaze away from the subject as if he were thinking. The confederate was trained to turn his gaze away at the same timing for every subject. The two conditions were different with regard to the confederate's behavior of turning the gaze:

    - *Condition 1 (EC)*: The confederate turns his gaze from the subject to the android as if he establishes eye contact with the android. As a result of this nonverbal behavior, the subject is assumed to feel as if a positive relationship has been established between the confederate and the android.
    - *Condition 2 (NEC)*: The confederate turns his gaze away from the android as if he averts the android's gaze.

    The motions of the android were triggered by the operator at a remote control station who monitored the interview. Under both conditions, the android nodded to greet the subject when he or she entered the experimental room, in order to make the subject feel as if it might be human. It looked at either the subject or the confederate during the interview, depending on which one was speaking.

5. *Measurement*: The questionnaire was designed to check the experimental setting and measure the subject's impression about the android. The most important questions were about the relationship between the subject and the confederate and the relationship between the confederate and the android, and they were evaluated on the basis of the confederate's eye contact with the android. We recorded the interviews by two video cameras and checked the behaviors of the subject and the android.

5.4.2 Results

1. *Manipulation check*: We analyzed both the video records and the questionnaire to reveal the subject's attitudes toward the android, the confederate, and their relationship.

   - Toward the android: There was no significant difference in the subject's behavior toward the android between the EC and NEC conditions – the duration of time in which the subject looked at the android (EC (M = 163.3, SD = 175.5), NEC (M = 264.5, SD = 322.2) [frame], t (26) = −1.00, n.s.), the duration of time and the number of times of the subject's eye contacts with the android ((EC (M = 0.29, SD = 0.47), NEC (M = 0.43, SD = 0.85), t (26) = −0.64, n.s.), (EC (M = 11.2, SD = 0.85), NEC (M = 18.9, SD = 40.2), t (26) = −0.54, n.s.)), and the number of times of the subject's nodding during the android's nodding (EC (M = 2.1, SD = 1.8), NEC (M = 2.9, SD = 2.2), t (26) = −1.02, n.s.).
   - Between the confederate and the android: We checked whether we successfully manipulated the subject's impression about the exchange between the confederate and the android. The scores for the question "Did you feel that the interviewee established eye contact with the IA recorder?" were significantly higher in the EC condition than in the NEC condition (t (26) = 3.65 × 10 − 8, $p < 0.01$), while the assumption of equal variances was not violated (F(13, 13) = 1.38, n.s). Note that two cases (one case in the EC and one case in the NEC condition) were judged as outliers and removed from the analysis, because the scores for this question were too far from the average scores in the conditions, i.e., larger than M +2 × SD or smaller than M −2 × SD.

     The scores for the question "Did you feel that the interviewee had a good impression toward the IA recorder?" were significantly higher in the EC condition than in the NEC condition (t (26) = 6.09 × 10 − 5, $p < 0.01$). Therefore, we concluded that we succeeded in influencing the subject's feelings about the relationship between the confederate and the android by controlling their eye contacts.
   - Toward the confederate: The scores for the question "Did you have a good impression toward the interviewee?" were higher than 4 (EC, M = 5.8; NEC, M = 5.2); 4 indicated a neutral feeling in the scale used in this experiment. There was no significant difference between them (t (26) = 0.12, n.s.), while the assumption of the equal variances was not violated (F(13, 13) = 2.57, $p = 0.10$). Therefore, the instruction and the dummy task worked to make the subjects evaluate the confederate as positive, and thus we regarded the subjects as having a feeling of a positive relationship with the confederate.

2. *Subjective effects*: We compared the average scores for the question (a) "Did you have a good impression about the IA recorder?" A two-tailed unpaired *t*-test revealed a significant difference between the conditions (t (26) = 0.0008, $p < 0.01$), while the assumption of equal variances was

not violated ($F(13, 13) = 2.58, p < 0.1$). The score in the EC condition was significantly higher than that in the NEC condition; therefore, it seemed that the subjects who had observed eye contacts between the confederate and the android had a more positive impression toward the android, in comparison with the subjects who had not.

5.4.3 *Conclusion*

The observed eye contact between the confederate and the android is likely to strengthen the subject's positive impression toward the android (question (a)). This result supports the hypothesis that one's impression toward a robot can be influenced by a person if he or she appears to have a nonverbal communication with it. In particular, one's preference for a robot is formed so as to mirror the other's attitude toward it, and the mirroring of attitudes is induced by observing the other's attentive behavior.

The result that a positive bias in preference formation is triggered by the other's attentive behavior is consistent with the prediction by Heider's balance theory. However, it is worth noting that preference formation is examined in another experiment that involves the supposition of a different set of positive or negative relationships among three agents, and the result of the experiment is biased and inconsistent with the prediction by Heider's theory. Since the manipulation of the relationships between the subjects and the confederate does not seem to be insufficient in this additional experiment, further examination (e.g., in more hostile relationships) is necessary for us to reveal the dynamics of preference formation and attentive behavior among three agents in more detail.

5.5 Conclusion

In this chapter, the elements of the dynamics of preference formation that are based on attentive behavior are analyzed by considering the nature of human cognition and sociality. These elements are categorized for the purpose of constructing social robots that can establish shared attention with humans. Three experiments are introduced to examine whether or how human social cognition is influenced in human-robot interaction. Since the experiments only give weak evidence, we should refine the robot's behavior and obtain a clearer view on the current results. This and future attempts are expected to promote the improvement of the mechanism of social robots in such a way that they will influence human cognition in human-robot interaction.

The study of the three ways of preference formation examined here, i.e., provoked attention, proxy response, and mirrored attention, has started just recently. Since these are newly found, new technology should be developed to utilize them. For example, in order to utilize the way of provoked attention for preference formation, a method for leading human attentive behavior in a desired direction is required. In addition, the mechanism of predicting human social behavior needs to be developed, so that a robot can know what action it should choose to lead human's attentive behavior.

The remaining two social ways of preference formation are open to multi-robot and multi-person situations. It is a growing field in human-robot interaction that studies and develops robots in multi-robot and multi-person situations. Interestingly, the cognitive processes underlying these ways of preference formation seem to be relevant to the development of telecommunication robot system, which is another active field in human-robot interaction. Since robots used in telecommunication should play the role of a physical proxy agent in communication, they should have the relevant characteristics of the teleoperator and make it easy to interpret and participate in what is going on at a distant place. Teleoperation can be regarded as a special type of multi-agent interaction that involves a teleoperator, a robot, and persons at a distant place. The reactions that it receives at the distant place should be substituted for those that happen near the teleoperator. The action that it produces at the distant place should be interpreted as the teleoperator's own action; in other words, it should induce mirrored attention. The phenomena to be utilized in face-to-face multi-agent interaction are the same as the one introduced in this chapter. Therefore, the studies of telecommunication with robots and multi-person interaction involving robots should help each other, and this will provide a chance to correctly generalize the phenomena of interest in each field and to utilize the effective methods for influencing humans that are found in each field.

Exercises

1. Pick an example of the way of measuring one's preference and then discuss its strong and weak points.
2. Discuss the strong and weak points of the three types of psychological experiments of human-robot interaction: using a robot with very humanlike appearance (such as android), using a robot with moderate humanlike appearance (such as M3-Synchy introduced in this chapter), and using a robot with less human likeness (such as Robovie-II or Paro).
3. Design a collective behavior of grouped robots to mediate the process of preference formation.

References

Breazeal, C.: Toward sociable robots. Robot. Auton. Syst. **42**(3–4), 167–175 (2003)
Heider, F.: The Psychology of Interpersonal Relations. Wiley & Sons, New York (1958)
Imai, M., Ono, T., Ishiguro, H.: Physical relation and expression: joint attention for human-robot interaction. IEEE Trans. Ind. Electron. **50**(4), 636–643 (2003)

Ishiguro, H., Minato, T.: Development of androids for studying on human-robot interaction. In: Proceedings of the 36th International Symposium on Robotics, TH3H1, Boston (2005)

Ishiguro, H.: Scientific issues concerning androids. Int. J. Robot. Res. **26**(1), 105–117 (2007)

James, W.: What is an emotion? Mind **9**(34), 188–205 (1884)

Kanda, T., Ishiguro, H., Imai, M., Ono, T.: Development and evaluation of interactive humanoid robots. Proc. IEEE **92**, 1839–1850 (2004)

Masuda, T., Ellsworth, P.C., Mesquita, B., Leu, J., Tanida, S., Van de Veerdonk, E.: Placing the face in context: cultural differences in the perception of facial emotion. J. Personality Social Psychol. **94**(3), 365–381 (2008)

Moore, C., Dunham, P. (eds.): Joint Attention: It's Origins and Role in Development. Lawrence Erlbaum Associates, Hillsdale (1995)

Robins, B., Dickerson, P., Stribling, P., Dautenhahn, K.: Robot-mediated joint attention in children with autism – a case study in robot-human interaction. Interact. Stud. **5**(2), 161–198 (2004)

Shimada, M., Yoshikawa, Y., Asada, M., Saiwaki, N., Ishiguro, H.: Effects of observing eye contact between a robot and another. Int. J. Soc. Robot. **3**(2), 143–154 (2011)

Shimojo, S., Simion, C., Shimojo, E., Scheier, C.: Gaze bias both reflects and influences preference. Nat. Neurosci. **6**, 1317–1322 (2003)

Sidner, C.L., Lee, C., Morency, L.-P., Forlines, C.: The effect of head-nod recognition in human-robot conversation. In: Proceedings of ACM/IEEE 1st Annual Conference on Human-Robot Interaction, pp. 290–296 (2006)

Watanabe, T., Danbara, R., Okubo, M.: Effects of a speech-driven embodied interactive actor "interactor" on talker's speech characteristics. In: Proceedings of the 12th IEEE International Workshop on Robot-Human Interactive Communication, pp. 211–216, California, USA (2003)

Yoshikawa, Y., Shinozawa, K., Ishiguro, H., Hagita, N., Miyamoto, T.: Responsive robot gaze to interaction partner. Robot. Sci. Syst. II, 287–293 (2006)

Yoshikawa, Y., Shinozawa, K., Ishiguro, H.: Social reflex hypothesis on blinking interaction. In: Proceedings of the 29th Annual Conference on the Cognitive Science Society, pp. 725–730. Nashville, Tennessee (2007)

Yoshikawa, Y., Yamamoto, S., Sumioka, H., Ishiguro, H., Asada, M.: Spiral response-cascade hypothesis – intrapersonal responding cascade in gaze interaction. In: Proceedings of the 3rd ACM/IEEE International Conference on Human-Robot Interaction, Amsterdam, Netherlands (2008)

Chapter 6
Communication for Social Robots

Takayuki Kanda and Takahiro Miyashita

Abstract This chapter overviews the studies in social robotics that deal with communication. First, we discuss how humans' natural communication is modeled into social robots. Second, we introduce a field study on how people and robots engage in communication in real-world environments.

Keywords Communication • Social robots • Computational model • Deictic communication

6.1 Introduction

For human beings, communication seems to be an easy task. We can perceive various kinds of information, process them, and communicate them with other people, but usually, we are not aware of what we do. When we plan to build a robot that can communicate with people in a "social" enough way, we realize how difficult it is to build such a robot and how complicated our internal process of natural and social communication with others is.

To date, psychology, cognitive science, or robotics provides no complete model of communication, and no robot is able to have humanlike communication. Nevertheless, recent research has started providing insights into the mechanism for communication and opened up the possibility that people can engage in social communication with robots. This chapter describes a couple of successful cases of *synthetic approaches* to understanding human communication.

Social communication involves nonverbal interactions and is context dependent. The first section of this chapter discusses the mechanism of social communication. The second section reports experiments in which people communicate with social robots. That is, even though the current capabilities of social robots are limited, they can achieve some degree of humanlike communication by using a synthetic approach.

T. Kanda (✉) • T. Miyashita
Intelligent Robotics and Communication Laboratories (IRC), Advanced Telecommunications Research Institute International (ATR), Keihanna Science City, Kyoto, Japan
e-mail: kanda@atr.jp

M. Kasaki et al. (eds.), *Cognitive Neuroscience Robotics A*,
DOI 10.1007/978-4-431-54595-8_6

6.2 Computational Model of Communication for Social Robots

What is substantially important for enabling social robots to have humanlike communication? Early research modeled communication as the process of accurately conveying information from a sender to a recipient. Models, such as the "code model," were built and developed for the purpose of representing this explicit flow of information.

However, it is now well known that various nonverbal elements are involved and given important roles in communication. Contrary to verbal exchange, nonverbal interaction is often *implicit*. In the field of human-robot interaction, many elements are under active research: for example, gestures (Breazeal et al. 2005; Nagai 2005; Scassellati 2002; Sidner et al. 2002), gaze (Mutlu et al. 2006; Nakano et al. 2003; Sidner et al. 2004), and timing (Kuzuoka et al. 2008; Shiwa et al. 2009; Yamamoto and Watanabe 2008).

In this section, we introduce two cases to investigate this nonverbal implicit interaction in more detail. Both cases are important for revealing the relevant aspects of human likeness in communication: simple exchange of information does not yield natural interaction, but extra nonverbal interaction is the key for humanlike natural interaction.

6.2.1 Natural Deictic Communication

6.2.1.1 Introduction

This study focuses on deictic communication. In casual communication, people often use reference terms in combination with pointing; for example, they say "look at *this*" while pointing to an object (this section is adopted from Sugiyama et al. (2007)). A simple view on this typical deictic communication treats it as the use of pointing gestures with reference terms. A previous study already made it possible for a communication robot to both understand what is pointing and itself point to an object with reference terms (Sugiyama et al. 2006).

However, this simple view only encompasses the elements pertaining to the interpretation of information, and it lacks a process of facilitating interaction. This study aims to reveal the importance of three major elements that facilitate deictic communication. The first element is "attention synchronization." When we point to an object, we pay attention to whether the listener's gaze is following our pointing gesture. Thus, if a communication robot is listening to a person who shows attention-drawing behavior, it needs to indicate that it is paying attention to the pointing gesture in order to facilitate the person's attention-drawing behavior.

The second element is "context": when we use a reference term such as "look at *this* (a white round box)," we omit its details, such as "white round box." This is relatively easy to understand if the context (in this example, the box) is established before the use of reference terms.

The third element is "believability." People sometimes feel that deictic communication is inaccurate (Sugiyama et al. 2005). For people to have a successful communication with a robot, they need to believe that the robot is capable of error correction; otherwise, people may hesitate to use deictic gestures/utterances toward a robot.

6.2.1.2 Naturalness in Deictic Communication

We have outlined five processes for natural deictic communication between robots and humans:

1. *Context focus*: The speaker and listener share the object of focus on the basis of the context of verbal communication.
2. *Attention synchronization*: The listener pays attention to the indicated object by synchronizing with the speaker's attention-drawing behavior.
3. *Object recognition*: The listener recognizes the indicated object on the basis of the speaker's attention-drawing behavior.
4. *Believability establishment*: The speaker corrects a recognition error of the listener when she notices it.
5. *Object indication*: The listener can change the role from listener to speaker and show an attention-drawing behavior to confirm that she has recognized the object.

We categorize object recognition and object indication processes as "interpretation processes" and remaining three processes as "facilitation processes." Detailed explanations of these processes are given in what follows.

6.2.1.2.1 Interpretation Process

When we talk about an object, we often point to it and apply a reference term to it, such as "this" and "that," to draw the listener's attention to the object. The use of a reference term and pointing can be dealt with in two processes:

- Object recognition

This is the process of interpreting an attention-drawing behavior on the basis of a pointing posture, a reference term, and object-property information in the utterance (Sugiyama et al. 2006).

- Object indication

This is the process of generating a robot's attention-drawing behavior with pointing and a reference term in order to confirm for the interacting person that it has recognized the object.

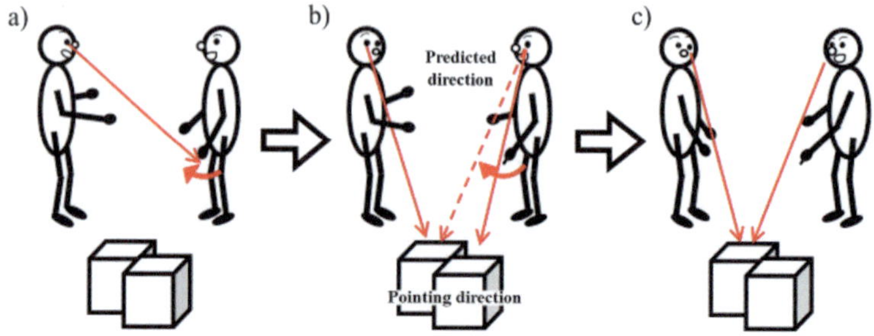

Fig. 6.1 The model of attention synchronization

6.2.1.2.2 Facilitation Process

However, the interpretation processes are not sufficient to achieve natural deictic communication. When we talk about an object, the speaker always monitors the listener's reaction to her behavior and evaluates whether the listener understands. At the same time, the listener grasps the intention of the speaker's deictic behavior and synchronizes with her gaze by following that behavior. That is, deictic communication is not a simple process where one person points to an object and another person passively interprets the pointing.

We believe that a robot should react to a human listener's behavior and simultaneously generate a situation where the human listener feels comfortable in accepting the robot's attention-drawing behavior. To build such a robot, we distinguish three processes: attention synchronization, context focus, and believability establishment.

- Attention synchronization

This process provides the interacting person with a feeling that the robot is attending to her pointing behavior. The flow of attention synchronization is shown in Fig. 6.1. It consists of three subprocesses:

(a) Reaction to the initiation of pointing

When the listener notices the initiation of the speaker's attention-drawing behavior, the listener immediately starts following it with her gaze.

(b) Prediction of indicated object

Next, the listener estimates the intention of the speaker's attention-drawing behavior and predicts what object the speaker intends to indicate. The listener often starts looking at the object before the speaker's pointing motion is finished.

(c) Attention sharing

Lastly, when the speaker finishes moving her hand for pointing, the speaker keeps the pointing gesture toward the pointed object and looks at it. At this time, the listener looks at the object. Thus, they share their attention toward the object when the speaker's pointing motion is completed.

- Context focus

This process sets a precondition for deictic communication. In interhuman communication, the speaker and the listener share the object of focus on the basis of the context of verbal communication before their deictic communication begins.

- Believability establishment

This process provides the robot "believability," that is, a user regards the robot as a believable entity in deictic interaction. Communication involves ambiguous expressions, and each participant in communication must believe that other participants can handle ambiguous expressions. If this kind of believability is not established, a person would hesitate to interact with the robot in a deictic way. In this research, we build believability into a robot by making it capable of error correction.

6.2.1.3 System Configuration

Utilizing the abovementioned processes, we have developed a communication robot that can engage in natural deictic communication with a human.

6.2.1.3.1 Hardware Configuration

This robot system was developed from a communication robot "Robovie" (Kanda et al. 2002), a motion-capturing system, and a microphone attached to a user. Robovie is 1.2 m tall and 0.5 m in diameter and has a humanlike upper body designed for communicating with humans. It has a head (3 DOF), eyes (2*2 DOF), and arms (4*2 DOF). With its 4-DOF arms, it can point with a gesture similar to human's pointing.

6.2.1.3.2 Software Configuration

The configuration of the developed robot system is shown in Fig. 6.2. The interpretation processes are achieved by the white-colored components. The facilitation processes are achieved by the orange colored components. Each process gets its input from either/both speech recognition or/and gesture recognition and controls and generates a behavior as its output. For speech recognition input, we used a speech recognition system (Ishi et al. 2008) and attached a microphone to human conversation partners in order to avoid noise. The speech recognition system can recognize reference terms, object-color information, and error-alerting words uttered by humans. Gesture recognition is done with a motion capture system. In gesture recognition, the system handles two types of processes: pointing posture recognition and pointing motion detection. How each process is implemented is described in detail in what follows.

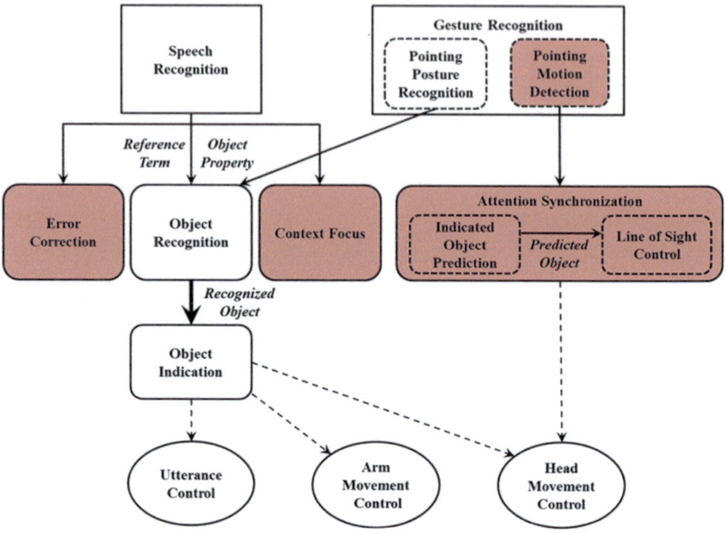

Fig. 6.2 System configuration

6.2.1.3.3 Interpretation Processes

The interpretation processes were developed on the basis of the three-layered model previously developed in Sugiyama et al. (2006). In object recognition, the system receives a reference term and object-property information from the speech recognition module and pointing gesture information from the gesture recognition module; they are important for identifying the indicated object. In object indication, the system generates an appropriate attention-drawing behavior, and it can confirm, using a deictic expression, that it has recognized the indicated object. The details are described in Sugiyama et al. (2006).

6.2.1.4 Facilitation Processes

This section describes the implementation of the facilitation processes into our robot; they are designed to facilitate natural deictic communication between the robot and humans. Each facilitation process is implemented as follows:

- Context focus

The "context focus" process is important for voice communication with humans. In the first stage of interaction in our experiment, the robot (R) and the human (H) have a short conversation:

R: May I help you?
H: Yes, please help me to put away these *round boxes*.
R: OK. Would you tell me the order of the *boxes* to put away?

In this case, the robot and the human can focus on the round boxes in the environment on the basis of the context described here. It is very important for both to be aware of specific objects indicated by the context. In this research, the robot could only handle the round boxes. Thus, this conversation is only for the human to mentally prepare for the objects in the context. It is left for future work to enable robots to distinguish among objects on the basis of contextual information.

- Attention synchronization

The process of attention synchronization is implemented as three subprocesses: pointing motion detection, indicated object prediction, and line of sight control. The important requirement for attention synchronization is that the robot starts looking at the indicated object quickly before the person finishes the pointing motion. Thus, the robot has to predict what is the indicated object from the pointing motion (Sugiyama et al. 2007).

- Error correction

The process of error correction establishes the believability of the robot. It is important for voice communication with humans. If the process of speech recognition detects an error-alerting word in a human utterance, this process is called on by the system. In this process, the robot asks the human to repeat her indication and deletes the latest recognized object from its memory.

6.2.1.5 Experiment

We have conducted an experiment to verify the effectiveness of the facilitation processes (Fig. 6.3).

Fig. 6.3 A scene of the experiment

6.2.1.5.1 Method

Experimental Environment: The experiment was conducted in a 3.5 × 3.5 m area at the center of a room. There were five cylindrical boxes. At the beginning of a session, each subject was asked to place these five boxes freely in the area where Robovie could point.
Subjects: Thirty university students (16 men, 14 women).
Conditions: We adopted 2 (facilitation process) × 2 (method of instruction) conditions as follows:

Facilitation process factors

(a) With-facilitation process

For error correction, subjects were instructed how to correct errors of the robot before the session; they could do so freely as they gave orders. For context focus, they were asked to mention the context when they ordered the robot. For example, they might say "please bring the boxes." During the session, the robot controlled its gazing direction and looked at the place the subject was going to point. This is based on the mechanism of attention synchronization, as we developed it. Note that the robot also looked at the boxes in the confirmation phase.

(a') Without-facilitation process

The subjects were given no instruction mentioned above.

Method of instruction factor

(b) Deictic method (pointing + reference term)

Subjects were asked to use pointing gesture, reference terms, and object-color information when they give orders to the robot.

(b') Symbolic method

Subjects were asked to read the numbers on the boxes when they give orders to the robot. A two-digit ID in 14-point font was attached to each box, and it was readable in a distance of 2 m.

This setup is intended to simulate the situation where there is some difficulty in finding the symbol that is used to identify an object. In everyday situations, for example, we might be told "please look at the book entitled *Communication Robots*"; then, we need to see the letters on the object and read them. This request taking is simulated using two-digit IDs.

The experiment was a within-subject design, and the order of experimental trials was counterbalanced (for these terms, see Chap. 7).

6.2.1.5.2 Procedures

Before the experiment, the subjects were instructed on how to interact with the robot, by both the deictic method (reference terms and pointing) and the symbolic method. After they had the instructions, the subjects experienced four sessions in all four conditions. In each session, they followed the following procedure three times:

1. They freely place the five boxes.
2. They talk about the context in which they are in.
3. They decide the order of the five boxes.
4. They indicate the five boxes one by one. The method of instruction is either (b) or (b'), depending on the experimental condition. For example, under (b) condition, a subject might point to the first box and say "this" and then point to the second box and say "that white one." The subject continues this until the fifth object.
5. After the subjects indicate the fifth box, Robovie repeats the order. The robot looks at each object and confirms its existence by utterance. For example, under (b) condition, Robovie uses a reference term while pointing to each of the five boxes.

After this process was repeated three times, the subjects filled out a questionnaire and specify their impressions of the interaction with the robot.

6.2.1.5.3 Measurement

We expected that the facilitation processes would make deictic communication with the robot more natural. In addition, we expected that due to the attention synchronization mechanism, the subjects would have a high feeling of sharing spatial information on the boxes with the robot. Thus, we used the following questionnaire to measure the subjects' impressions. The subjects answered each question on a 1-to-7 scale, where 1 stands for the lowest and 7 stands for the highest evaluation.

Naturalness: Your feelings of naturalness of the conversation
Sharing information: Your feelings of sharing information with the robot
To compare the deictic method and the symbolic method, we were interested in quickness and correctness (Sugiyama et al. 2006). Thus, we also measured them in the questionnaire:
Quickness: Quickness of your indication of the boxes
Correctness: Correctness of the robot's identification of the box you have indicated
Understanding: The robot's understanding of the indication

In addition, we measured the system's performance in terms of the accuracy of recognizing the subject's indication in both the deictic method and symbolic method. The main cause of error comes from a failure in speech recognition. Since speech recognition sometimes fails and the deictic method utilizes multimodal inputs, we expected the deictic method to provide better performance.

Performance: The rate at which the robot system correctly identified the object indicated by the subjects

Table 6.1 Performance

	Without facilitation (%)	With facilitation (%)
Deictic	98.4	98.8
Symbolic	95.9	95.0

a)

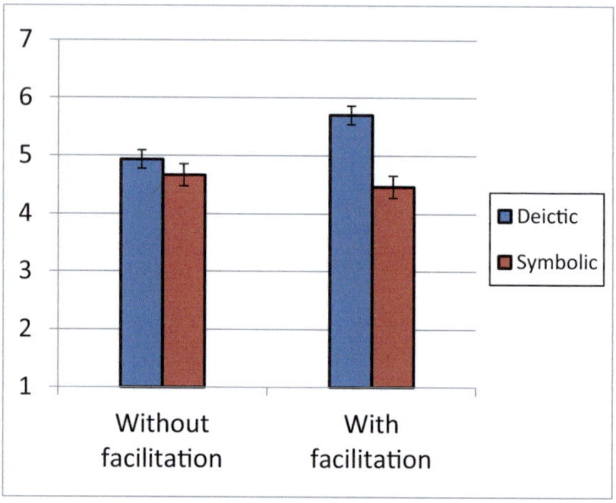

b)

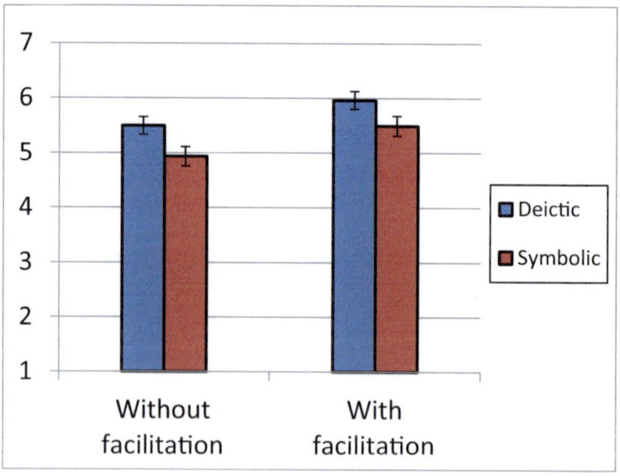

Fig. 6.4 Results for facilitation process. (**a**) Naturalness and (**b**) sharing information

6.2.1.6 Results

Table 6.1 shows the system's performance. Figures 6.4 and 6.5 show the results of the questionnaire. In order to analyze the results, we conducted repeated measures ANCOVA (analysis of covariance); it is an extension of repeated measures ANOVA (analysis of variance) to include covariance. There were two within-subject factors,

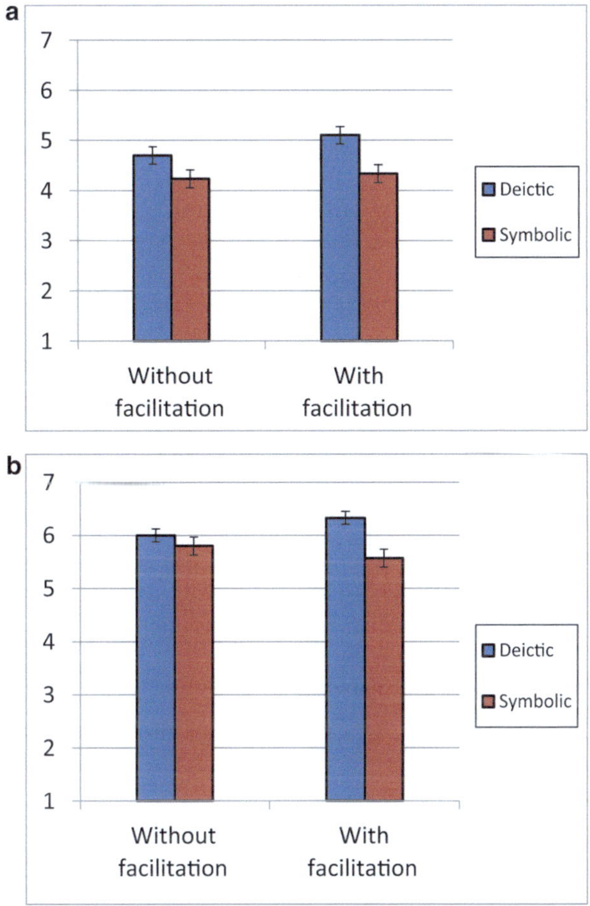

Fig. 6.5 Results for method of instruction. (**a**) Quickness and (**b**) robot's understanding

"facilitation process" (with-facilitation process or without-facilitation process) and "method of instruction" (deictic method or symbolic method). In addition, we used "performance" (the rate at which the robot system correctly identified the object) for the covariance, since some of the impressions, such as the impression of the robot's understanding, could be affected by the robot's real performance rather than by the conditions.

Naturalness (Fig. 6.4a): There was a significant difference in method of instruction ($F(1,86)=8.142$, $p<0.01$) and the interaction between the two factors ($F(1,86)=4.988$, $p<0.05$). Analysis of the interaction revealed that there were a significant difference in method of instruction under the with-facilitation process condition ($p<0.01$) and a significant difference between the facilitation processes under the deictic communication condition ($p<0.01$). Thus, the facilitation process contributed to the feeling of natural interaction when the subjects interacted with the robot in deictic communication.

Sharing information (Fig. 6.4b): There were significant differences in facilitation process ($F(1,86)=7.121$, $p<0.01$). There were nearly significant differences in method of instruction ($F(1.86)=3.782$, $p=0.055$) and no significant difference in the interaction between the two factors. Thus, the subjects felt that the robot with a facilitation process was better at sharing information than the robot without it. In addition, it was suggested that deictic communication could contribute to the feeling of sharing information better than symbolic communication could.

Quickness (Fig. 6.5a): There were significant differences related to the instruction method ($F(1.86)=8.100$, $p<0.01$). There was no significant difference in facilitation process, the interaction between the two factors, and performance. Thus, the subjects felt that the deictic method is quicker than the symbolic method.

Correctness: There was no significant difference in facilitation process, instruction method, and the interaction between the two factors. There was nearly a significant difference in performance ($F(1.86)=2.978$, $p=0.088$).

Understanding (Fig. 6.5b): There was a significant difference in performance ($F(1,86)=6.453$, $p<0.05$) and no significant difference in the other factors. Thus, subjects distinguished the robot's performance and evaluated it as the robot's understanding.

Performance (Table 6.1): For performance, we conducted a within-subject design ANOVA, and it indicates a significant difference in method of instruction ($F(1,86)=16.300$, $p<0.01$). There was no significant difference in facilitation process.

6.2.1.7 Conclusion

This study has investigated three facilitation processes in deictic communication: attention synchronization, context focus, and believability establishment. The experiment result reveals that the facilitation processes make deictic communication natural. The subjects report that the robot feels to them to share information with the robot when the facilitation processes are used. In addition, the comparison of the symbolic-command condition and the proposed method reveals that deictic communication is more effective for recognizing the indicated objects due to multimodal processing.

We take this study to illuminate the depth of human communication. The interpretation processes are sufficient merely for information exchange; however, the extra processes, i.e., the facilitation processes, are important for people to have deictic communication with the robot.

6.2.2 *Providing Route Directions*

6.2.2.1 Introduction

We believe that a promising application of social robots is to provide route directions (Kanda et al. 2009; Shiomi et al. 2008) (this section is adopted from Okuno et al. (2009)). Hence, this study investigates a way to enable a social robot to provide route directions.

How can a robot give good route directions? If the destination is within a visible distance, the answer might be intuitive: the robot would say "the shop is over there" and point to the shop. However, since the destination is often not visible, the robot needs to utter several sentences with gestures. We design our robot's behavior in such a way to enable the listener to intuitively understand the information provided by the robot. This study illustrates how we integrate three important factors of giving directions—*utterances*, *gestures*, and *timing*—so that the robot can provide good route directions.

6.2.2.2 Modeling of Robot's Route Directions

We modeled the generation process of a robot's route directions and divided it into three models: *utterances*, *gestures*, and *timing*.

6.2.2.2.1 Utterance

In developing a model of utterances, we reviewed the literature and decided to rely on the work of Daniel et al. (2003): they concluded that a route description should contain minimal information that is neither too short to avoid ambiguity nor too detailed. From this standpoint, they proposed a "skeletal description," which consists of a series of sentences. The key idea is that each sentence contains a landmark and an action. An action is an instruction about walking behavior, such as "go straight," "turn left," or "turn right." A landmark is an easy-to-find building in the environment where people are instructed to take an action, such as a bank, a post office, or a library. Thus, a sentence should be something like "turn left at the bank." Following the idea of skeletal description, the robot uses such a sentence to provide information about how to reach the destination.

6.2.2.2.2 Gesture

Since people also use gestures to give route directions, the next question to consider is what kinds of robot gestures can help people understand route directions. We have reviewed the literature (e.g., Allen 2003; Kendon 2004; Striegnitz et al. 2005) and classified gestures often used in route directions into four types: deictic gesture, orienting body direction, iconic gesture (expressing landmarks), and beat gesture.

For our experiment, we decided to focus on deictic gestures. We did not use iconic and beat gestures, because it was unclear whether they would work positively or negatively; we intended to reveal the usefulness of the most promising gesture, the deictic gesture, and left other gestures to future study. The deictic gesture is used to point to the absolute direction.

6.2.2.2.3 Timing

The robot pauses after it utters a sentence. A model of *timing* purports to decide the duration of a pause. We take two different approaches for modeling the timing: from the speaker's perspective (i.e., natural timing for speaking) and from the listener's perspective (i.e., time needed to understand). Later, we experimentally decide which one enables a robot to give better route directions.

- Speaker's timing model

A model of human speaker behavior is commonly used to create naturalness in computers; instances of naturalness are speech synthesis, CG agents, etc. Once the model is well developed, a robot could naturally provide route directions at the timing modeled on the human speaker's timing.

To model the timing of speakers, we measured their pause durations. We asked six students in our laboratory to read sentences with a "skeletal description." The description consisted of six sentences. Students looked at the description for a minute and then read it out to a listener in a natural manner. The listener was standing in front of the speaker but did not provide any particular reaction (e.g., nodding) to the speaker. We measured the pause duration after the speaker read each sentence.

Figure 6.6 shows the average pause duration of the six speakers. *Pi* represents a pause after *i*th sentence. The *P1* to *P5* averages range between 0.42 and 0.66 [s]. On the basis of the "speaker" model, we built from this experiment the robot used the measured *P1* to *P5* as pause durations in providing route directions.

- Listener's timing model

An alternative approach is to model the time that people take to understand utterances. It is a cognitively demanding task to listen route directions, and it requires the comprehension of spatial relationships and the memorization of routes and landmarks. Thus, it would increase the listener's understanding if a robot takes enough time after each utterance.

We did an experiment to model how people use time to understand a route direction (Okuno et al. 2009). Eight university students participated in this experiment. We measured the time span between the end of the robot's utterance and the participant's report that she understood it. This measurement was repeated four

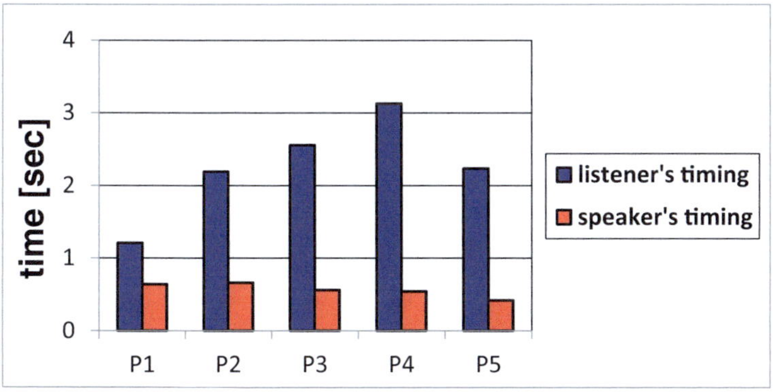

Fig. 6.6 Timing model

times for each participant with different route directions. Figure 6.6 (upper line) shows the average time span listeners took to understand an utterance. It is the average gained from eight participants with four trials. On average, *P1* to *P5* ranged between 1.03 and 3.07 [s]. On the basis of this "listener" model, the robot used the measured *P1* to *P5* as pause durations in providing route directions. These values are quite long in comparison to those found in previous speech synthesis studies.

6.2.2.3 Experiment

We conducted an experiment to measure an effect of gestures and find a better model of timing. In addition, we compared the robot's task performance with human's and evaluated the effectiveness of our model.

6.2.2.3.1 Method

- Participants

Twenty-one native Japanese speakers (14 males and 7 females, all undergraduate students) participated in our experiment for which they were paid.

- Settings

We used Robovie. The experiment was conducted in a 3 × 3 m space. An A0 size picture of a street in a town was presented on the wall. A speaker (a human or a robot) and a listener (a participant) stood near the picture (Fig. 6.7).

- Conditions

There were two conditions with different gestures and timings. In addition, different people provided route directions.

Fig. 6.7 A scene of the experiment

Conditions for the robot

The robot gave directions on the basis of the "skeletal descriptions," (Daniel et al. 2003) and each description contained six sentences.

(a) Gesture

With: The robot performed deictic gestures.
Without: The robot did not perform any gestures; it expressed no arm movements.

(b) Timing

Speaker: The robot uttered sentences on the basis of the speaker model.
Listener: The robot uttered sentences on the basis of the listener model.

Condition with a human speaker

Two human speakers gave route directions. They were from our laboratories but did not know the purpose of the experiment. One gave detailed route directions, and the other gave directions that resembled the skeletal descriptions.

- Procedure

The experiment was a within-subject design, and the order of sessions was counterbalanced. At the first session, participants were instructed to imagine that they were on the way to a popular restaurant and get lost in an unfamiliar area. They were instructed to ask for a direction and then to draw it on a piece of paper after listening to the route direction provided by the speaker. They were also told that they had to indicate landmarks on the way to the restaurant and what they do to reach the restaurant. Participants were positioned in the "listener" position, and the robot/human speaker stood in the "speaker" position.

The participants had four sessions, i.e., repeated the experiment four times with different route directions. After a route direction was provided at each session, they drew a map and completed a questionnaire. Note that we prohibited the participants from asking for a route direction more than once; thus, a route direction was given only once in each session.

6.2.2.3.2 Measurement

We conducted two types of evaluation.

Correctness We asked the participants to draw a map and indicate how to get to the destination. We counted the number of correct actions and landmarks on the map. Consequently, the score ranged from 0 to 8.

We also asked the participants to evaluate their impressions about the following items on a 1-to-7 scale, where 1 stands for the lowest and 7 for the highest evaluation.

Easiness: How easy/difficult was it to understand the route direction?
Naturalness: How natural/unnatural was the route direction?

6.2.2.3.3 Hypothesis and Predictions

Earlier studies show that human gestures are useful for human listeners, and this suggests that the robot's gestures would help human listeners to understand route directions. Figure 6.6 shows that listeners need more time to understand a route direction than the time speakers take to pause, and this suggests that the speaker model would not offer enough time for listeners to understand. Thus, we made the following predictions:

1. When the participants listen to the route directions with gestures, correctness scores and easiness ratings will be higher than when they listen without gesture.
2. When the participants listen to the route directions at the timing controlled by the listener model, correctness scores and easiness ratings will be higher when they listen at the timing controlled by the speaker model.

6.2.2.4 Results

6.2.2.4.1 Verification of Predictions

For the correctness scores (Fig. 6.8), a two-way repeated measures analysis of variance (ANOVA) was conducted with two within-subject factors, *gesture* and *timing* A significant main effect was revealed in both the gesture factor ($F(1,20) = 16.055$, $p < 0.005$) and the timing factor ($F(1,20) = 6.757$, $p < 0.05$), but no significance was found in the interaction within these factors ($F(1,20) = .323$, $p = 0.576$).

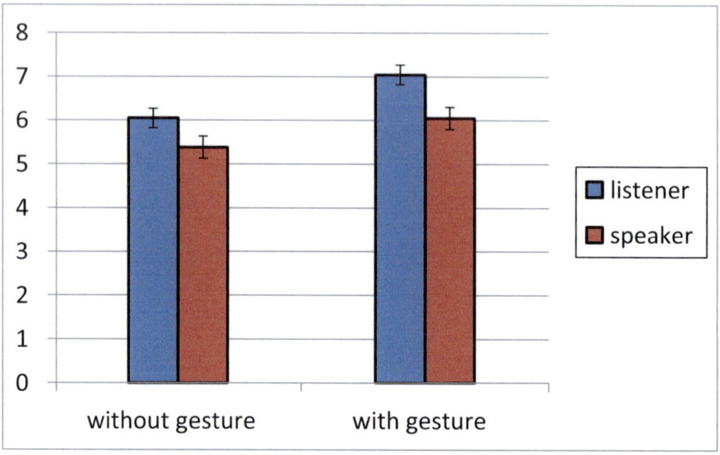

Fig. 6.8 Experimental results of correctness scores

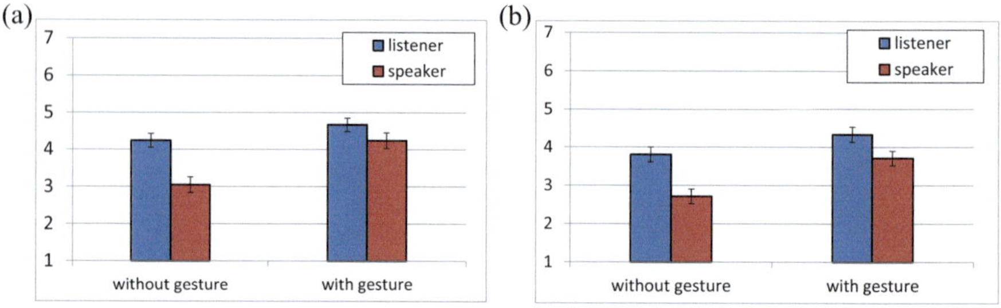

Fig. 6.9 Experimental results of subjective evaluations. (**a**) Easiness and (**b**) naturalness

For the easiness ratings (Fig. 6.9a), a two-way repeated measures ANOVA with two within-subject factors, *gesture* and *timing*, was conducted. A significant main effect was revealed in both the gesture factor ($F(1,20) = 13.945, p < 0.005$) and the timing factor ($F(1,20) = 12.105, p < 0.005$). The interaction within these factors was also significant ($F(1,20) = 6.448, p < 0.05$).

There was a simple main effect in the gestures in the speaker model ($F(1,20) = 28.027, p < 0.001$), while no significant main effect was observed in the gestures in the listener model ($F(1,20) = 2.077, p = 0.165$). There was a simple main effect in the timing in the without-gesture condition ($F(1,20) = 30.941, p < 0.001$), but no significant main effect was observed in the timing in the with-gesture condition ($F(1,20) = 1.709, p = 0.206$).

Overall, predictions 1 and 2 are supported by this experiment.

Fig. 6.10 Comparing route directions of human and robot

6.2.2.4.2 Comparison of Naturalness Ratings

A two-way repeated measures ANOVA was conducted with two within-subject factors, *gesture* and *timing*, for the result of naturalness (Fig. 6.9b). A significant main effect was revealed in both the gesture factor ($F(1,20) = 8.928$, $p < 0.01$) and the timing factor ($F(1,20) = 12.308$, $p < 0.005$). The interaction within these factors was also significant ($F(1,20) = 5.525$, $p < 0.05$). We analyzed the simple main effects in the interaction within these factors. It was almost significant in the gestures in the listener model ($F(1,20) = 3.467$, $p = 0.077$) and significant in the gestures in the speaker model ($F(1,20) = 14.000$, $p < 0.005$), in the timing in the with-gesture condition ($F(1,20) = 5.972$, $p < 0.05$), and in the without-gesture condition ($F(1,20) = 15.838$, $p < 0.005$). Overall, participants rated naturalness higher for the robot when it uttered with gestures or at the timing controlled by the listener model. The combination of the two factors did not improve the naturalness rating twice, as in the case of the easiness ratings.

6.2.2.4.3 Comparison with the Route Directions Given by Humans

To analyze how well the robot gave route directions, we compared its route directions and humans' (Fig. 6.10). The robot condition contains the "with-gesture" and "listener model" conditions; this combination was found in the experiment above to be the best design for the robot's route directions.

A one-way repeated measures ANOVA was conducted with one within-subject factor, *speaker* (human or robot). The robot performed better in the correctness scores ($F(1,41) = 7.920$, $p < 0.01$), there is no significant difference in the easiness ratings ($F(1,41) = .067$, $p = 0.796$), and humans gave better impression in the naturalness ratings ($F(1,41) = 10.348$, $p < 0.005$).

It is interesting that the robot's providing route directions led to higher correctness scores; however, the participants regarded the robot's direction giving as not very easy and more unnatural than the human's direction giving. This seems to suggest that the robot's direction giving could be improved. For example, its utterance is too simple in comparison with humans' utterance. Humans used more conversational fillers and conjunctions. In addition, it could be not easy to compare naturalness between the robot and humans, as people might apply different criteria for naturalness toward the robot and humans.

6.2.2.5 Summary

The effect of the robot's gestures is demonstrated in the experiment on route directions where deictic gestures are used in parallel with utterances. That is, even though utterances provide enough information for a listener to understand a direction, gestures add more information to utterances. Furthermore, the importance of timing is highlighted. People prefer a pause duration modeled on the time they need to understand, even though the pause duration is significantly longer than usual. This long pause duration could be interpreted as unnatural. But interestingly, the experimental result has revealed that the participants rated it as more natural; our interpretation of this result is that the participants were so busy understanding the route directions that they did not feel unnaturalness in this long silence.

6.3 Case Studies

6.3.1 Introduction

Recent progress in robotics has enabled us to start developing humanoid robots that interact with people and support their everyday activities (this section is adopted from Kanda et al. (2009)). We believe that humanoid robots are suitable for communicating with humans. Previous studies have demonstrated the merits of the physical presence of robots for providing information. Moreover, their humanlike bodies enable them to perform natural gaze motion and deictic gestures.

These features of humanoid robots might allow them to perform communicative tasks in human society, such as giving directions or acting as a tour guide in exhibitions. Since we are not yet sure what communicative tasks robots will engage in the future, many researchers have started conducting field trials to explore possible tasks. This is an important research activity, since few real social robots are working in the everyday environments. Case studies in the real environments enable us to envision the future scenes of human-robot interaction, and this is helpful for solving the problems concerning the social acceptance of robots in everyday life.

The study introduced in this section focuses on an information-providing task in an everyday environment, that is, in a shopping mall. Compared with a school or a museum, people in a shopping mall are often busy, seeking some specific item, and have no special interest in robotics technology; the environment of a shopping mall is challenging. We aim to answer the following questions:

- Can a robot function to perform an information-providing task in an open public environment, such as a shopping mall?
- Can a robot influence people's everyday activities, such as shopping?
- Can a robot elicit spontaneous interaction from ordinary people repeatedly?

6.3.2 Design

In this subsection, we state how we designed the robot's roles, how we realized them in the system framework, and how we considered them while creating the robot's interactive behaviors. This subsection provides an opportunity to consider the design process of a social robot that works in the real world.

6.3.2.1 Contemplating Robot Roles

Many other devices than robots, such as maps and large screens, provide information. Robots have unique features relevant to its physical existence, its interactivity, and its capability for personal communication. We define three roles of a guide robot in a shopping mall as follows:

6.3.2.1.1 Role 1: Guiding

The size of a shopping mall continues to grow larger and larger. People sometimes get lost in a mall and ask for a direction. Even though a mall has maps, many people still prefer to ask for a help. Some information is not shown on a map; thus, people ask questions like "where can I buy an umbrella?" A robot has unique features not shared by a map and other devices: it physically exists, it is colocated with people, and it is equipped with humanlike body properties. Thus, a robot can naturally teach a direction by pointing in a humanlike manner, look in the same direction as a person, and use reference terms, such as "this way."

6.3.2.1.2 Role 2: Building Rapport

Since a robot is a representative of the mall, it needs to be friendly and comfortable to customers. In addition, since a mall is a place that people repeatedly visit, a robot needs to naturally repeat interaction with the same person; thus, a function that builds rapport with each customer is useful.

In the future, a robot may need to have a function of customer relationship management. This was currently done by humans: for example, the shopkeeper at a small shop remembers the "regulars" and molds communication with each of them. The shopkeeper may be kind to particular good customers. In places where the number of customers is too big to manage, information systems assume this role in part, such as mileage services of airplane companies, point systems of credit cards, and online shopping services. However, these information systems do not provide natural personalized communication as humans do; we believe that a future robot is able to provide natural communication and personalized service for individual customers and to develop a good relationship or rapport with them.

6.3.2.1.3 Role 3: Advertisements

Advertisement is important for shopping malls. For instance, posters and signs are placed everywhere in a mall. Recently, information technologies are being used for advertisement as well. We believe that a robot can be a powerful tool for this purpose. Since a robot is novel, it can attract people's attention and direct their interest to the information it provides.

6.3.2.2 System Design

What role a robot can take is limited by its recognition and action capabilities, which are largely constrained by its hardware and infrastructure. Thus, first, we should consider the system design of a robot (hardware and infrastructure). We need to explore a promising combination of hardware and infrastructure. Some researchers are studying a stand-alone robot that is capable of sensing, decision making, and acting. By contrast, some researchers are focusing on a combination of robots, ubiquitous sensors, and humans. We have chosen the latter strategy, known as a "network robot system" (Sanfeliu et al. 2008), in which a robot's sensing and its decision processes are supported by ubiquitous sensors and a human operator, respectively.

The most important component of our system for users is a robot. It provides information in a natural way with its capabilities for speaking and making gestures. For this reason, users can concentrate on the robot standing in front of them.

In a network robot system, most of the intelligent processing is done apart from the robot. Sensing is mainly done by ubiquitous sensors. There are three important sensing elements in our system: *position estimation*, *person identification*, and *speech recognition*. For *position estimation*, we use floor sensors that accurately and simultaneously identify the positions of multiple people. This could also be done with other techniques, such as distance sensors. For *person identification*, we employ a passive-type radio-frequency identification (RFID) tag that always provides accurate identification. This tag requires intentional user contact with an

RFID reader; since passive-type RFIDs have been widely used for train tickets in Japan, we consider this unproblematic.

We use a human operator for *speech recognition* and *decision making*. If the way of providing information is instable and awkward, it causes people to be disappointed. The quality of current speech recognition technology remains far from useful. For instance, a speech recognition system prepared for noisy environments, which performs 92.5 % word accuracy in 75 dBA noise (Ishi et al. 2008), has only 21.3 % accuracy in a real environment (Shiomi et al. 2008). This shows difficulties in recognizing everyday utterances: changes of voice volume among people and/or within the same person and the unpredictability of noise in a real environment. A speech recognition program causes many recognition errors, and hence robots often have to ask for elucidation.

6.3.2.3 Behavior Design

6.3.2.3.1 General Design

We set two basic policies for designing the robot's interaction. First, it takes the communication initiative and introduces itself as a guide robot. It gives a direction and then provides information in response to user requests. This way, customers clearly understand that the robot is engaged in route guidance. Second, the robot makes utterances and performs other behaviors in an affective manner, not in a reactive manner. The robot engages in humanlike greetings, reports its "experience" with products in shops, and tries to establish a good relationship (or rapport) with customers. This is very different from the master-slave-type communication where a robot prompts a user to provide a command.

6.3.2.3.2 Guiding Behavior

There are two types of behaviors prepared for guiding: *route guidance* and *recommendation*. The former is a behavior in which the robot teaches how to get to a destination with utterances and gestures, as shown in Fig. 6.1. The robot points to the first direction and says "please go that way" with an appropriate reference term chosen by the attention-drawing model (introduced in the earlier part of this chapter). It goes on to say: "After that, you will see the shop on your right." Since the robot knows all of the mall's shops and facilities (toilets, exits, parking, etc.), it can teach about 134 destinations.

In addition, in situations where a user has not decided where to go, we designed *recommendation* behaviors: the robot makes suggestions on restaurants and shops. For example, when a user asks for a good restaurant, the robot starts a dialogue by asking a few questions, such as "What kind of food would you like?" and chooses a restaurant to recommend.

6.3.2.3.3 Building Rapport Behavior

If a person wears an RFID tag, the robot starts to build a rapport with that person. The robot performs three types of behaviors. First, the robot performs self-disclosure behavior. For example, the robot mentions its favorite food, "I like *Takoyaki*," and its experiences, such as "this is my second day working in this mall."

Second, because we have found in our previous study (Kanda et al. 2004) that people appreciate it if robots call their names, the robot calls a person's registered name and greets with, e.g., the utterance of "Hello, Mr. Yamada." In addition, the robot uses the memory of the previous dialogue to inform that the robot remembers the person.

Third, the robot changes behaviors according to friendliness. It gradually changes its behavior to show a more and more friendly attitude toward a person if she repeatedly visits.

6.3.2.3.4 Advertising Behavior

The robot is also intended to provide advertisements about shops and products in a manner that resembles people's *words of mouth*. When the robot starts a conversation with a customer, it starts with a greeting and then engages in a *word-of-mouth* behavior as a form of casual chat. It affectively reports its pretended experiences with products in shops. For example, the robot may say "yesterday, I ate a crêpe in the food court. It was nice and very moist. I was surprised!" or "The beef stew *omu-raisu* [omelet rice] at Bombardier Jr. was good and spicy. The egg was really soft, too, which was also very good." We implemented five topics per day and changed the topics every day so that regular shoppers did not get bored with this behavior.

6.3.3 System

Figure 6.11 shows an overview of the system. The hardware used for the system was Robovie, the same robot as used in our early experiments. We invited customers to join a field trial and gave them an RFID tag for person identification. A passive-type RFID tag was embedded in a cellular phone strap. Customers were instructed to place their tag on the reader attached to the chest of the robot for identification and to interact with the robot. Person detection was performed by six floor sensor units around the robot that covered a 2×2 m area. The robot identified a person by an RFID tag reader and tracked his/her position by floor sensors.

According to the information from sensors, a behavior selector chooses an interactive behavior on the basis of the pre-implemented rules called "episode rules" and the memory of the past dialogues with this person. Interactive behavior is implemented with *situated modules* ("behavior" in this paper) and *episode rules* (Kanda et al. 2004). Each situated module controls the robot's utterances, gestures, and nonverbal behaviors in reaction to a person's action. For example, when the

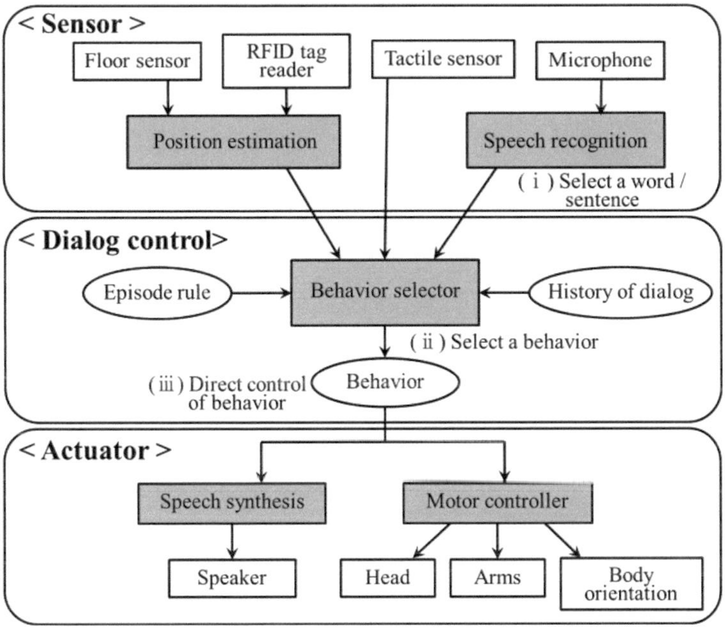

Fig. 6.11 System configuration

robot attempts to shake hands by saying "let's shake hands," it waits for input from a tactile sensor to react to the person's handshake. Each behavior usually lasts for about 5–15 s. In total, 1759 *situated modules* and 1015 *episode rules* are implemented.

The robot system is designed to operate without an operator; however, since speech recognition often fails in a noisy environment, there are many unexpected conversational situations where the robot cannot correctly respond to requests. Thus, in this study, we used the Wizard of Oz (WOZ) method with a human operator to supplement these weaknesses of autonomous robots. The italic letters in Fig. 6.11 indicate the elements controlled by the operator. We asked the operator to minimize the amount of operations. Except for speech recognition, the operator only helped the robot when an operation was truly needed (further details of the system are reported in Kanda et al. (2010)).

6.3.4 Field Trial

6.3.4.1 Procedure

The field trial took place at a shopping mall with three floors. It had approximately 150 stores and a large supermarket. The robot was placed in a main corridor of the

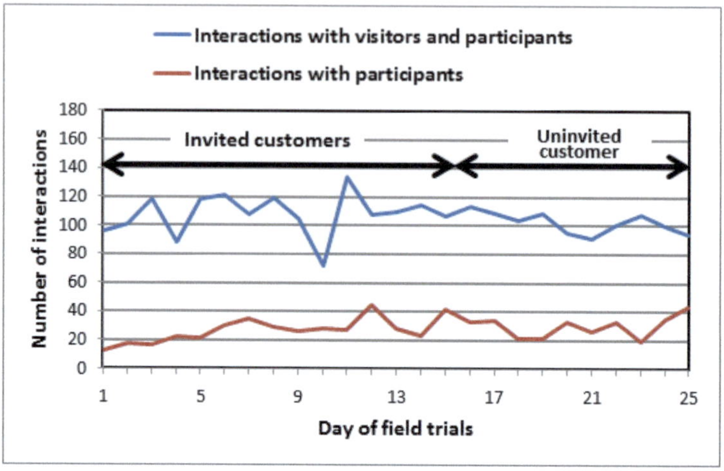

Fig. 6.12 Daily visitors and participants

mall in weekday afternoons (1:00–5:00 pm) for 5 weeks (July 23 to August 31, 2007, except for a busy week in the middle of August).

The robot was open to all visitors. Those who signed up for the field trial (participants) received a passive-type RFID embedded in a cell phone strap. We recruited these participants by two methods: (1) flyers distributed to residents around the mall and (2) on-site invitation during the first 3 weeks; our staff approached visitors who seemed interested in the robot. The participants filled out a consent form at the beginning and a questionnaire at the end the field trial.

6.3.4.2 Results

6.3.4.2.1 Overall Transition of Interactions

Figure 6.12 shows the number of interactions the robot engaged in with the trial participants. Here, one interaction with a participant is assumed to end when the robot says goodbye. Figure 6.13 shows the interaction scenes. During the first 3 weeks, our staff invited visitors to the field trial and asked them to have interaction with the robot. From the fourth week, our staff stood near the robot for safety reasons. As the graph shows, the number of interacting persons did not differ in the 5-week period. Multiple persons interacted with the robot at the same time (an average of 1.9 persons per interaction).

Three hundred and thirty-two participants signed up for the field trial and received RFID tags; 37 participants did not interact with the robot at all, 170 participants visited once, 75 participants visited twice, 38 visited 3 times, and 26

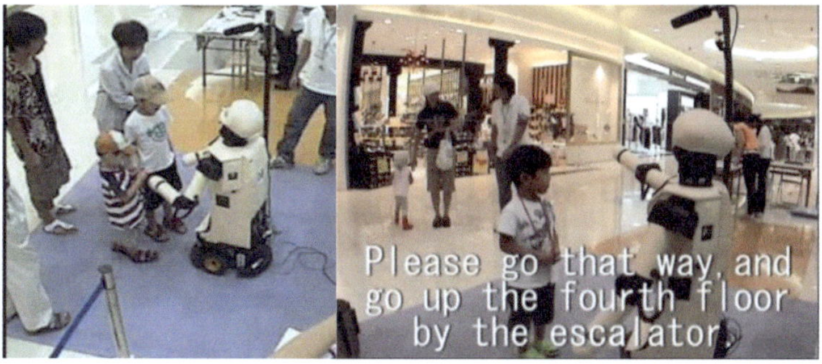

Fig. 6.13 Interaction scenes

visited 4 times; the remaining 23 participants visited 5–18 times. On average, each participant interacted 2.1 times with the robot, indicating that they did not repeat interaction very much. One obvious shortage was the trial duration; since many nonparticipant visitors waited in line to interact with the robot, some participants reported that they hesitated to interact with the robot. Figure 6.12 shows the number of the participants who interacted each day, with an average of 28.0 persons per day.

6.3.4.2.2 Perception of Participants

When the field trial finished, we mailed a questionnaire to the 332 participants and received 235 answers. All items were evaluated on a 1-to-7 point scale where 7 represents the most positive, 4 represents neutral, and 1 represents the most negative.

Impression of robot: The questionnaire included items: "Intention of use" (studied in Heerink et al. (2008)), "(the degree of) Interest," "Familiarity," and "Intelligence." Their respective scores were 5.0, 4.9, 4.9, and 5.1 (S.Ds. were 1.3, 1.4, 1.4, and 1.4).

Route guidance: The answers to the question about the adequacy of route direction had 5.3 points on average (S.D. was 1.3).

Providing information: The answers to the questions about the usefulness and interest in the information provided by the robot had 4.6 and 4.7 points on average, respectively (S.Ds. were 1.4 and 1.3). Moreover, 99 out of the 235 participants reported that they visited a shop mentioned by the robot, and 63 participants bought something on the basis of the information provided by the robot.

Building rapport: The answer to the question about the degree of felt familiarity had 4.6 point on average (S.D. was 1.5).

Overall, the robot was positively felt and received by the participants.

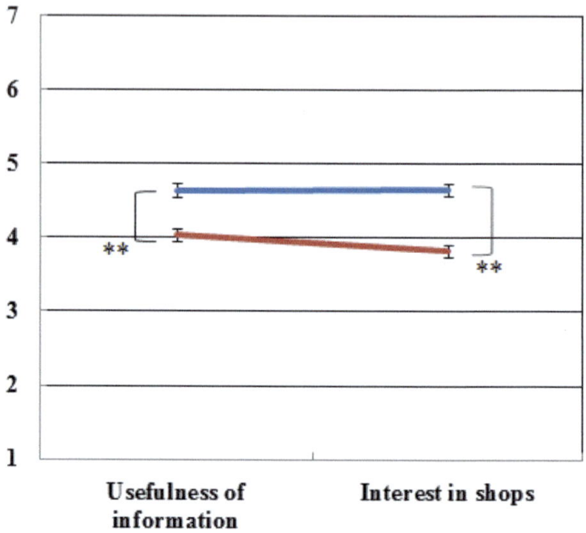

Fig. 6.14 Comparison of impressions of the robot and display

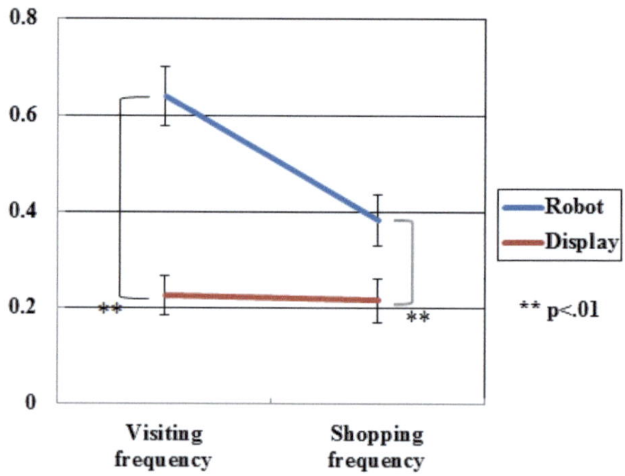

Fig. 6.15 Comparison of influences from the robot and display

6.3.4.2.3 Comparison with an Information Display

We asked the participants how often they were influenced by information displays in the same mall. In more detail, they were asked to answer the following: "Usefulness of information provided by the display/robot," "Interest in shops mentioned by the display/robot," "Visiting frequency triggered by the display/robot," and "Shopping frequency triggered by the display/robot." The order of the questions about the display and robot was counterbalanced.

Figures 6.14 and 6.15 show the comparison result. There were significant differences ($F(1229) = 40.96$, 69.52, 36, 19, and 7.66, $p < 0.01$ for all four items). Thus, the robot provided more useful information and elicited more shopping from the participants than the display did.

6.3.5 *Summary*

We have developed a robot that can provide services, such as route guidance and other information, in a shopping mall. A 5-week field trial was conducted in a shopping mall. The results indicate that customers accept the robot; they have positive impressions of the robot and are influenced by the information it provides. The robot has performed well in the information-providing task and successfully influenced people's shopping activities.

Acknowledgment The authors of this chapter give their thanks to the coauthors of the original papers from which parts of this chapter are adopted: Osamu Sugiyama, Yusuke Okuno, Michita Imai, Masahiro Shiomi, Zenta Miyashita, Hiroshi Ishiguro, and Norihiro Hagita.

Exercises

Choose one of the two essay topics, and write an essay on it. Make sure to answer the questions in case of Essay 1.

- Essay 1 (from Sect. 6.2):

  Imagine a scene in which a communication involves implicit nonverbal interaction. Then, consider how a computational model (i.e., make a robot capable of dealing with the scene) can realize communication in the scene.

- What techniques are required (sensing, computation, actuation, etc.)?
- How can the computational model be developed?

- Essay 2 (from Sect. 6.3):

- The Three Laws of Robotics are written by Isaac Asimov in 1942: they are about *human safety*, *obedience to humans*, and *self-defense*. Given the progress in social robots (e.g., an example of field study shown in Sect. 6.3), what would be the Three Laws for Social Robots, if they were codified now?

References

Allen, G.L.: Gestures accompanying verbal route directions: do they point to a new avenue for examining spatial representations? Spat. Cogn. Comput. **3**(4), 259–268 (2003)

Breazeal, C., Kidd, C.D., Thomaz, A.L, Hoffman, G., Berlin, M.: Effects of nonverbal communication on efficiency and robustness in human-robot teamwork. Paper presented at the IEEE/RSJ International Conference on Intelligent Robots and Systems (IROS2005) (2005)

Daniel, M.-P., Tom, A., Manghi, E., Denis, M.: Testing the value of route directions through navigational performance. Spat. Cogn. Comput. **3**(4), 269–289 (2003)

Heerink, M., Kröse, B., Wielinga, B., Evers, V.: Enjoyment, intention to use and actual use of a conversational robot by elderly people. Paper presented at the ACM/IEEE International Conference on Human-Robot Interaction (HRI2008) (2008)

Ishi, C.T., Matsuda, S., Kanda, T., Jitsuhiro, T., Ishiguro, H., Nakamura, S., Hagita, N.: A robust speech recognition system for communication robots in noisy environments. IEEE. Trans. Robot. **24**(3), 759–763 (2008)

Kanda, T., Ishiguro, H., Ono, T., Imai, M., Nakatsu, R.: Development and evaluation of an interactive humanoid robot "Robovie". Paper presented at the IEEE International Conference on Robotics and Automation (ICRA2002) (2002)

Kanda, T., Hirano, T., Eaton, D., Ishiguro, H.: Interactive robots as social partners and peer tutors for children: A field trial. Hum-Comput. Interact. **19**(1&2), 61–84 (2004a)

Kanda, T., Ishiguro, H., Imai, M., Ono, T.: Development and evaluation of interactive humanoid robots. Proc. IEEE. **92**(11), 1839–1850 (2004b)

Kanda, T., Shiomi, M., Miyashita, Z., Ishiguro, H., Hagita, N.: An affective guide robot in a shopping mall. Paper presented at the ACM/IEEE International Conference on Human-Robot Interaction (HRI2009) (2009)

Kanda, T., Shiomi, M., Miyashita, Z., Ishiguro, H., Hagita, N.: A communication robot in a shopping mall. IEEE. Trans. Robot. **26**(5), 897–913 (2010)

Kendon, A.: Gesture: Visible Action as Utterance. Cambridge University Press, Cambridge/New York (2004)

Kuzuoka, H., Pitsch, K., Suzuki, Y., Kawaguchi, I., Yamazaki, K., Yamazaki, A., Kuno, Y., Luff, P., Heath, C.: Effect of restarts and pauses on achieving a state of mutual orientation between a human and a robot. Paper presented at the ACM Conference on Computer-supported cooperative work (CSCW2008). http://dl.acm.org/citation.cfm?id=1460563.1460594&coll=DL&dl=GUIDE&CFID=240417221&CFTOKEN=57240792 (2008)

Mutlu, B., Forlizzi, J., Hodgins, J.: A storytelling robot: modeling and evaluation of human-like gaze behavior. Paper presented at the IEEE-RAS International Conference on Humanoid Robots (Humanoids'06) (2006)

Nagai, Y.: Learning to comprehend deictic gestures in robots and human infants. Paper presented at the IEEE International Workshop on Robot and Human Interactive Communication (RO-MAN2005) (2005)

Nakano, Y.I., Reinstein, G., Stocky, T., Cassell, J.: Towards a model of face-to-face grounding. Paper presented at the Annual Meeting of the Association for Computational Linguistics (ACL 2003) (2003)

Okuno, Y., Kanda, T., Imai, M., Ishiguro, H., Hagita, N.: Providing route directions: design of robot's utterance, gesture, and timing. Paper presented at the ACM/IEEE International Conference on Human-Robot Interaction (HRI2009) (2009)

Sanfeliu, A., Hagita, N., Saffiotti, A.: Special issue: network robot systems. Robot. Auton. Syst. **56**(10), 793–797 (2008)

Scassellati, B.: Theory of mind for a humanoid robot. Auton. Robot. **12**(1), 13–24 (2002)

Shiomi, M., Sakamoto, D., Kanda, T., Ishi, C.T., Ishiguro, H., Hagita, N.: A semi-autonomous communication robot – a field trial at a train station. Paper presented at the ACM/IEEE International Conference on Human-Robot Interaction (HRI2008) (2008)

Shiwa, T., Kanda, T., Imai, M., Ishiguro, H., Hagita, N.: How quickly should a communication robot respond? Delaying strategies and habituation effects. Int. J. Soc. Robot. **1**(2), 141–155 (2009)

Sidner, C.L., Dzikovska, M.: Hosting activities: experience with and future directions for a robot agent host. Paper presented at the International Conference on Intelligent User Interfaces (IUI 2002) (2002)

Sidner, C.L., Kidd, C.D., Lee, C., Lesh, N.: Where to look: a study of human-robot engagement. Paper presented at the International Conference on Intelligent User Interfaces (IUI 2004) (2004)

Striegnitz, K., Tepper, P., Lovett, A., Cassell, J.: Knowledge representation for generating locating gestures in route directions. Paper presented at the Workshop on Spatial Language and Dialogue (5th Workshop on Language and Space) (2005)

Sugiyama, O., Kanda, T., Imai, M., Ishiguro, H., Hagita, N.: Three-layered draw-attention model for the humanoid robots with gestures and verbal cues. Paper presented at the IEEE/RSJ International Conference on Intelligent Robots and Systems (IROS2005) (2005)
Sugiyama, O., Kanda, T., Imai, M., Ishiguro, H., Hagita, N.: Three-layer model for generation and recognition of attention-drawing behavior. Paper presented at the IEEE/RSJ International Conference on Intelligent Robots and Systems (IROS2006) (2006)
Sugiyama, O., Kanda, T., Imai, M., Ishiguro, H., Hagita, N.: Natural deictic communication with humanoid robots. Paper presented at the IEEE/RSJ International Conference on Intelligent Robots and Systems (IROS2007) (2007)
Yamamoto, M., Watanabe, T.: Effects of time lag of utterances to communicative actions on embodied interaction with robot and CG character. Int. J. Hum-Comput. Interact. **24**(1), 87–107 (2008)

Chapter 7
System Evaluation and User Interfaces

Hideyuki Nakanishi

Abstract It is essential for developing a useful system to evaluate its user interface. The interface should be evaluated by potential users instead of experts, if unbiased results are to be obtained. It is more recommendable to observe users' reactions to an actual working prototype of the interface than to simply ask them about what kind of interface is preferable. The observation in a laboratory room can produce more general and scientific results than that in a real-world situation does. Thus, this chapter describes how to conduct a laboratory study. It collects users' reactions to the prototype of a user interface in a controlled situation. There are many items to consider in order to conduct a laboratory study: goals, hypotheses, factors, conditions, experimental design, tasks, subjects, data collection, and analysis. Each of these items is explained in detail. An example of system evaluation that is conducted in the past study on telepresence robots is discussed.

Keywords Human-computer interaction (HCI) • Usability • User study • Iterative development • Order effect • Between-subjects experiment • Ceiling effect • Human-robot interaction (HRI) • Questionnaire • ANOVA

7.1 Why Is System Evaluation Necessary?

System evaluation that focuses on the quality of user interfaces is important for any system that needs user intervention. The degree of user intervention varies with kinds of systems. Computer systems have been the typical target of evaluation, since they are usually very complex. Human-computer interaction (HCI) is the research field that deals with human factors in computer systems. Various kinds of user interfaces have been created in this field with the aim of improving the ease of controlling complex systems.

Complex systems do not necessarily require complex control. The degree of user intervention depends on the complexity and autonomy of the system. Complex systems that are highly autonomous need only a small degree of user intervention;

H. Nakanishi (✉)
Graduate School of Engineering, Osaka University, Suita, Osaka, Japan
e-mail: nakanishi@ams.eng.osaka-u.ac.jp

M. Kasaki et al. (eds.), *Cognitive Neuroscience Robotics A*,
DOI 10.1007/978-4-431-54595-8_7

users only need to turn on them. Neither simple systems nor autonomous systems require complex control. However, it is not true that those systems do not have the problem of bad user interface. The problem may exist even in the minimum intervention of simply turning on the system. It may be difficult to find the button to turn on the system. And it may be difficult to understand how to press the button. Thus, system evaluation is important even for systems that require only simple control.

Doors, water faucets, and light switches are typical examples of simple systems that are often difficult to use. A knob is attached to both sides of the door in Japan. So, it is difficult to determine whether the door should be pulled or pushed. The shape of some water faucet is a horizontal or vertical lever, and it is easy to distinguish between the levers that can be turned to the left or right and the levers that can be tilted up or down. However, the directions for faucet knobs to turn for on and off are often very confusing. Compared with water faucets, the directions for light switches to turn are easier to recognize, since they are usually equipped with some kind of mark or signal that indicates their current state. However, in a large room, such as a classroom, many lights are attached to the ceiling, and so it is hard to guess the correspondence between lights and switches.

The usefulness of a system depends on its utility and usability. Many people tend to evaluate a system in terms of utility. For example, many people think that a system is useful if it has a lot of functions, is high in performance, or resists breakdown. However, usability is as important as utility, since bad usability hampers good utility. Bad user interfaces do not allow users to use good functions. Even though a system is very high in performance, the system is of no value if users cannot control it. Therefore, it is important that system evaluation focuses on the quality of user interfaces.

7.2 Who Should Evaluate a System?

In the evaluation of a system, the most important question is who should evaluate it. Several different kinds of people are involved in developing the user interface of a system. Designers determine the look and feel of the interface, engineers provide technologies that are necessary to implement the interface, and psychologists analyze the cognitive aspect of the interface. Which kind of person is most suitable for the evaluation? The answer is that none of them is suitable, since they have their own bias. Designers believe that their experience, intuition, and ideas are effective in designing interfaces. Engineers optimistically expect that excellent technologies can always satisfy users. Psychologists are more interested in human nature than systems and are sometimes motivated by their doubts about the state-of-the-art technologies; so they tend to focus on human adaptability and reject the new features of systems.

There are two types of methods for evaluating a system: methods in which experts evaluate the system and methods in which users evaluate the system. The examples of the former type are the keystroke level model, cognitive walk-through, and heuristic evaluation. The keystroke level model is a method in which an expert tries to predict the time required for a person to command a system to do something. The command is defined as a series of key inputs and mouse operations. To estimate the length of time, the expert simply sums up the time of each key input and each mouse operation. If a user interface makes the time shorter, it is regarded as a better one. The cognitive walk-through is a method in which an expert tries to imagine how users who have no knowledge about a system learn how to use the system. Since the process of such learning is usually trial and error, the expert deliberates how users can face errors and judges whether they can recover from them. An error from which it seems hard to recover indicates the need for improvement. Heuristic evaluation is a method in which a team of experts tries to find discrepancies between the design of the user interface of a system and the guidelines that should be followed by the design. Experts form a team because a single expert cannot find all discrepancies. If there are a sufficient number of experts, this method is effective for improving the user interface.

As described above, system evaluation by experts is a simulation of a user's reaction to the system. Therefore, there is always an obvious difference between the simulated user's reaction and the actual user's reaction. This chapter describes system evaluation by users, which is called the "user study." In a user study, the users or evaluators of a system use the system for some task through its user interface. The experimenters observe what the users do, ask the users how they feel about the system, and analyze the interaction between the users and the system.

7.3 When Should a System Be Evaluated?

System evaluation is often regarded simply as a ritual to confirm that the system actually works. This tendency becomes strong when the developer plays the role of an evaluator. In many cases, the development proceeds on the basis of the waterfall model, in which a system is designed, implemented, and tested sequentially. The model assumes that the design of a system does not change after the system is designed, and the implementation of a system does not change after the system is implemented. Thus, a user study is conducted after most of the development is finished. The redevelopment of the system is costly and tends to be avoided even if the results of the study indicate that the user interface of the system needs a serious improvement. Since the quality of a user interface can easily become low, system evaluation after development is undesirable.

It is ideal to develop a system on the basis of the results of system evaluation. If this is achieved, the quality of a user interface becomes high without cost for redevelopment. Of course, system evaluation before development is impossible in principle, because a user study requires a prototype system that works to the extent to which users can use it. However, there are several methods for evaluating a system on the basis of user's reactions without any working prototype.

One method that can evaluate a system before development is called "requirements analysis." In this method, the developer asks users to detail the specifications of a system that satisfy them. The users have to guess what kinds of functions are necessary for the system to help their tasks. The specifications obtained by this method are tentative, because it is difficult even for experts to determine the specifications of a future system that currently does not exist. Thus, the method can be used only for designing the first version of a prototype system.

Another method for system evaluation before development is called "paper prototyping." This method uses a prototype that simply consists of illustrations on paper. Such a prototype is not interactive by itself. To show a user the interactive behaviors of a system, an experimenter leads the user to see a different illustration according to the user's virtual input to the system. This method enables the experimenter to test many variations of a user interface without making a real prototype, but cannot replace a user study with the real prototype, since it is inherently difficult to reproduce the interactive behaviors of a system. To simulate the interactive behaviors realistically, the experimenter has to draw a lot of illustrations. Even if an enormous number of illustrations are prepared, it is difficult for this method to provide the user with the same experience as a real prototype does. Thus, the method can be used to check the appearance of a user interface but is unsuitable for checking its interactive behaviors.

As described above, there are limitations in the system evaluation that is conducted before development. However, the system evaluation that is conducted after development is undesirable. Therefore, system evaluation should be included in the process of development. The process of development including system evaluation is called the "iterative development"; it is the opposite of the waterfall model in which system evaluation is the last part of the process of development. An iterative development is a sequence of multiple small waterfall processes, each of which consists of the design, implementation, and test of a system. If problems are found in the test phase of a certain small waterfall process, the design and implementation phases improve the system in order to solve these problems. Thus, system evaluation is interleaved with the whole process of development.

In each small waterfall process of an iterative development, the design and implementation phases become much shorter than those of the process that is based on the normal waterfall model. Therefore, to develop a system on the basis of the iterative development model, the developer has to be able to prepare a prototype system in a very short time. There are two kinds of prototypes that can be prepared easily. One is a vertical prototype: only a few of all the functions are actually implemented in it, and it can work only for a limited task. This kind of prototype is effective for evaluating the utility of some specific functions of a system but is not effective for checking the usability of the overall system. The other is a horizontal prototype: no function of a system is implemented in it, but the user interface of the whole system is constructed to enable interaction between the system and users. The system's responses to the users are simulated. The horizontal prototype is useful for iterative development.

7.4 Where Can a System Be Evaluated?

Every system is supposed to be used in some real-world situations. For example, automatic teller machines are used in a bank, and ticket vending machines are used in a station. Do these machines need to be evaluated in real-world situations? It may not be necessary to evaluate the machines in real-world situations, because the usability of their user interfaces can be tested in a laboratory room. The test in a laboratory is called the "laboratory study." The validity of the layout of buttons, the method of using colors, and the labels on menu items seem to be related solely to human cognitive capability and independent of real-world situations. This seeming is partially true, and some laws can explain why it is true, e.g., the Fitts's law and Hick's law. The Fitts's law predicts the amount of time a person takes to use some input device (e.g., a mouse) to point to some target (e.g., an icon). According to the Fitts's law, the length of time is proportional to the binary logarithm of the distance between the target and the current pointing position that is divided by the size of the target. The Hick's law predicts the amount of time a person takes to choose an item among menu items. According to the Hick's law, the length of time is proportional to the binary logarithm of the number of menu items. These laws have been examined by many laboratory studies.

This chapter describes how to conduct a laboratory study because it is an effective tool for system evaluation. However, there are limitations in laboratory study, since there is a gap between a laboratory room and real-world situations. A laboratory room used for the system evaluation that focuses on the quality of user interfaces is sometimes called a "usability lab." In general, a usability lab consists of a test room and an observation room, and they are connected by a half mirror or a video connection. In the test room, a person who is usually paid some money for participating in the experiment operates the system to accomplish a task that is given by the experimenter. In the observation room, the experimenter looks at the mirror or video monitor to analyze how the user operates the system. Obviously, this situation is very different from real-world situations, since the user, task, and situation are artificially prepared.

Systems can be tested more naturally in the real world than in laboratory studies. A test in the real world is called a "field study," in which users and tasks are real. The problem of field studies is the difficulty of controlling experiments. In a laboratory study, it is easy to control experiments: collecting participants who have the same attributions (e.g., age, knowledge, and experience), giving the same task to all the participants, and preventing accidents that would affect the results of the experiments. Thanks to this ease of control, it is possible to obtain statistical results and general findings that can be applied to various systems. In a field study, on-site people use a system for their own purposes in a chaotic environment. For example, in a situation where passengers use a ticket vending machine in a station, its user interface does not include all the factors that change the time a passenger takes to buy a ticket. The length of time also depends on whether the passenger is in a hurry, what kind of ticket the passenger is trying to buy, and how much the surrounding noisy environment draws the passenger's attention. Since it is usually difficult to obtain statistical results in a field study, its results tend to be anecdotal. The findings from it tend to be related to a specific real-world situation that is used for the study. Thus, there is a trade-off between naturalness and generality.

7.5 How to Evaluate a System?

In summary, it is important to evaluate the user interface of a system, the user interface should be tested by users, an actual working prototype of the system interface should be used in the user study, and user studies are often conducted in a laboratory room. The rest of this chapter describes how to conduct a laboratory study for evaluating the user interface of a system. There are many things to consider in order to conduct a study: goals, hypotheses, factors, conditions, experimental design, tasks, subjects, data collection, and analysis. First, you have to define the goal of study, i.e., what you want to clarify. If there is no goal, there is no need to conduct the study. Since the definition of a goal influences the rest of the study, you should sufficiently deliberate the goal. Next, you have to establish hypotheses according to the definition of the goal. Each hypothesis is a simple explanation that can be examined by the study. The goal must be achieved by examining the hypotheses. Each hypothesis usually explains that some difference in user interface of the system changes a user's reaction to the system. The difference becomes a factor. To confirm whether the factor is really able to change a user's reaction, you need to prepare the conditions for measuring the reaction. The conditions differ from each other because they involve different factors. After preparing the conditions, you have to choose the experimental design, i.e., the method for assigning people who participate in the study to the conditions. In each condition, an experimenter asks the participants to do some task. You can simply choose a typical way of using the system as a task but that is not always a good choice. A good task is one that can reveal differences between the conditions. You also have to collect subjects, i.e., participants in the study. You should determine what types of people are suitable as subject according to the kind of task and the manner of data collection. Finally, you have to determine the method for analyzing the collected data. The rest of this section details each of the items described above. The description of each item is followed by a corresponding description of the example of system evaluation that is conducted in the past study on telepresence robots (Nakanishi et al. 2008).

7.5.1 Goal

The goal of a system evaluation is what you want to clarify. You may want to clarify many aspects of the system, but you should focus on a few vitally important aspects, since many subjects are necessary to clarify just one aspect. If you try to clarify many aspects at once, you may not be able to collect a sufficient number of subjects. An insufficient number of subjects are likely to lead to a vague result. From a scientific point of view, a small number of clear results are much better than a large number of vague results. The relevant question here is what kinds of aspects you should focus on. This question is inherently difficult to answer because the kinds of aspects you should focus on heavily depend on the actual system. But it is possible to say what kinds of aspects you should not focus on. There are two kinds of aspects that should be avoided. One is trivial aspects that do not strongly affect the value of the system. For example, the traditional criteria for evaluating usability, i.e., learnability, efficiency, and memorability, seem to be unsuitable for evaluating pet robots. Of course, they are usually equipped with some user interface for configuration, and the interface can be evaluated by the traditional criteria. But the capability of socially interacting with a user is their major value, and this social capability cannot be evaluated appropriately by the traditional criteria. Obviously, you should focus on social capability rather than the traditional criteria when evaluating pet robots. The other kinds of aspects you should not focus on are those that are specific to the prototype used in the study. This rule is not relevant to applied research, because its purpose is the improvement of some commercial products. However, the rule is vital to basic research, because it tries to obtain and advance scientific knowledge. The clarification of the aspects specific to the prototype enables the improvement of the prototype but does not lead to any general knowledge. In basic research, a prototype of a system is just an experimental device for producing some general knowledge, and the improvement of the prototype itself is not very important. Suppose that you compare two systems. It is not enough to say which system is better. You need to generalize the difference between them and show a way to apply the results of the comparison to other systems of a similar kind. So, you should focus on the aspects of the systems that are common to systems of similar kinds.

Telepresence Robots' Example Telepresence robots enable virtual face-to-face conversations among geographically distributed people. These robots have been studied for more than 10 years (Ishiguro and Trivedi 1999; Jouppi 2002; Paulos and Canny 2001) and become commercially available several years ago. Research prototypes and commercial products of telepresence robots differ in form but have similar structures. All of them are equipped with microphones and speakers for vocal conversation and with a camera that allows an operator to observe the on-site situation. In addition, they are usually equipped with a display that shows the operator's face. All of these devices are mounted on the robotic base that can be controlled remotely by the operator. In short, telepresence robots are videoconferencing terminals that are capable of moving in a physical space. Normal videoconferencing terminals cannot move in a physical space as humans do. To reproduce face-to-face conversations, videoconferencing terminals may need to be able to move around. This seems to have been the motivation for researching and developing telepresence robots. Thus, the goal of system evaluation is defined as the clarification of the psychological effects caused by telepresence robots that are movable videoconferencing terminals.

7.5.2 Hypothesis

The goal of system evaluation described above cannot be directly examined by experiments. So, you need to convert the goal into a set of hypotheses. A hypothesis should be a form that explains the relationship between factors (independent variables) and data that is collected in a study (dependent variables). In a laboratory study that evaluates the user interface of a system, the relevant factor is a difference in design of user interface, and collected data are the subjects' reactions to the system. A hypothesis is accepted if the predicted relationship between the different designs of user interface and the changes in the subjects' reaction is actually observed in experiments. You can approach the goal by accepting or rejecting the proposed hypotheses. The problem here is that it is sometimes difficult to make hypotheses from the goal of study, since it is usually abstract and vague. If this is the case, certain knowledge helps to make hypotheses. The most reliable knowledge is produced by well-confirmed theories, i.e., theories that have been examined in many past studies. For example, psychological or cognitive neuroscientific theories may be useful for constructing hypotheses and evaluating user interfaces. Your and other researchers' past work, if it is closely related to the goal of system evaluation, is also useful, though it might be incorrect and thus less reliable than well-confirmed theories. However, it frequently happens that you cannot find any relevant well-confirmed theory and have no choice but to rely on your or other researchers' past work. You may even need to rely on ordinary common knowledge, when you cannot draw on a sufficient amount of scientific knowledge.

Telepresence Robots' Example A past study shows that stereoscopic images strengthen the presence of a remote conversation partner in video conference (Prussog et al. 1994). This means that binocular parallax strengthens the presence of a partner. Motion parallax, which is another depth cue, may cause the same effect. The movements of most telepresence robots are a combination of right-left rotation and forward-backward movement. The forward-backward movement of the camera mounted on the robot produces motion parallax. Thus, an example of hypothesis made from this observation is that a stronger presence of remote conversation partner is produced when the robot's movement includes forward-backward movement, even if the robot does not rotate at all. Another past study suggests that the predictive feed-forward model explains how we can distinguish between self-generated and externally generated actions (Vogeley and Fink 2003). While user-controlled movement of a robot is predictive, automatically generated movement is not. Therefore, it may be thought that a user-controlled movement of the robot would be recognized as a movement of the user's body. Thus, another example of hypothesis is that user-controlled movement of a robot strengthens the presence of a remote conversation partner, but automatically generated movement of a robot does not. The two hypotheses are illustrated in Figs. 7.1 and 7.2.

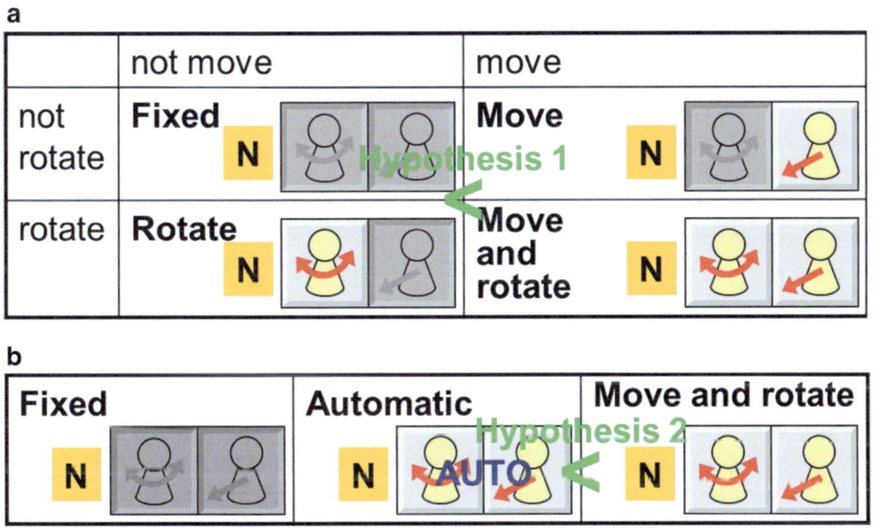

Fig. 7.1 Hypotheses, factors, conditions, and experimental design. (**a**) Comparison 1: 2 × 2 conditions in a between-subjects design. (**b**) Comparison 2: 3 conditions in a between-subjects design

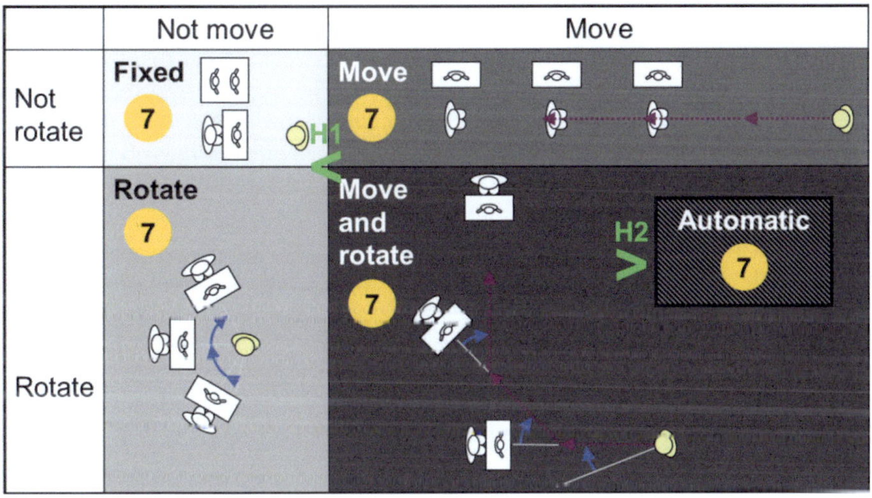

Fig. 7.2 Task and subjects

7.5.3 *Factors*

A hypothesis inherently includes the definition of a factor, since it's an explanation of the relationship between the factor and its influence on user experience; user experience is perception that results from the use of a system. Thus, it is easy to retrieve factors from hypotheses if they are correctly constructed. Factors should be deliberately defined, since they determine what the study can clarify and what kind of user interface prototype you need for the study. The most important principle for defining factors is that factors must be simple, since it is difficult for a complex factor, i.e., a combination of simpler factors, to produce clear results. Suppose that you want to clarify what kind of appearance is suitable for pet robots. If you straightforwardly choose the appearance of pet robots as a factor of the study, the results become ambiguous. This is because the factor is likely to be a combination of simpler factors (e.g., size and shape). For example, when you compare a small cat robot with a large dog robot and find that the former is more attractive than the latter, there are four ways of interpreting this result: small robots are more attractive, cat robots are more attractive, the combination of small size and catty shape makes robots attractive, and the combination of large size and doggy shape makes robots unattractive. This ambiguity stems from the fact that there are actually two factors, i.e., size and shape. You can easily interpret the result in the wrong way if you are unaware of either factor. Just comparing the two pet robots does not lead to any substantial conclusion, and it is almost impossible to know which of the four interpretations is correct, since the number of conditions, two, does not match the number of possible designs of the appearance, four. The next subsection details this issue.

Telepresence Robots' Example The first hypothesis is that a stronger presence of the remote conversation partner is produced when the robot's movements include a forward-backward movement, even if the robot does not rotate at all. This hypothesis includes two factors: the presence or absence of the robot's forward-backward movement and its right-left rotation. The second hypothesis is that user-controlled movement of the robot strengthens the presence of the remote conversation partner, but automatically generated movement of the robot does not. This hypothesis includes one factor: the robot is controlled by the subjects, moves automatically, or does not move at all. In Table Fig. 7.1a, the movement factor is represented as columns, and the rotation factor is represented as rows. In Table Fig. 7.1b, the factor of the second hypothesis is represented as columns.

7.5.4 Conditions

Each factor has several levels that correspond to its variations. For example, if a factor is the color of a button and there are three variations, red, green, and blue, then the number of levels is three. The number of conditions you need to consider equals the number of all combinations of each factor's levels. In the pet robots' example described in Sect. 7.5.3, there are two factors and each factor has two levels. So you need to consider four conditions: small cat, small dog, large cat, and large dog conditions. If you want to compare three sizes, small, medium, and large, then you need to consider six conditions that also include the medium cat and dog conditions. As you may notice, an increase in number of factors causes a combinatorial explosion in number of conditions, which makes it difficult to conduct the study. If you sharpen your focus on the system and keep the hypotheses simple, the number of factors and conditions remains feasible. It is recommended that you consider the conditions that cover all combinations of each factor's levels, but you can relax this requirement if you think that some of the combinations do not need to be examined in experiments for the following reasons. First, the combination may be a practically meaningless design for the user interface of the system. For example, you may think that the large cat condition can be omitted, since large cats are not as popular as large dogs. Second, a combination may be obviously superior or inferior to other combinations; it is unnecessary to compare a combination with other combinations. For example, you may predict that the large dog condition is rated as most unattractive because even people who are not afraid of real large dogs would be afraid of large dog robots. In this case, you can omit the large dog condition. Note that a decision to omit conditions might be wrong. The deletion of conditions sometimes leads to a reduction in number of factors. For example, if you decide to delete the large cat condition and compare only the small cat, small dog, and large dog conditions, then the study includes only one factor that has three levels. This factor seems to be as complex as the factor that is discussed in Sect. 7.5.3. However, the comparison of the three conditions can produce clearer results than the comparison of just two conditions, i.e., the small cat and large dog conditions; the result that the small cat condition is more attractive than the small dog condition implies the superiority of cat robots to dog robots, and the result that the small dog condition is more attractive than the large dog condition implies the superiority of small robots to large robots.

Telepresence Robots' Example The examination of the first hypothesis needs to compare four conditions to test two factors: the move and rotate, move, rotate, and fixed conditions. In the move and rotate condition, the robot moves and rotates. In the move condition, the robot only moves. In the rotate condition, the robot only rotates. In the fixed condition, the robot does not move, and so the subject does not need to control it. The examination of the second hypothesis needs to compare three conditions to test one factor: the move and rotate, automatic, and fixed conditions. In the automatic condition, the robot is actually controlled by the experimenter who precisely imitates the movements of the robot controlled by the subject in the move and rotate condition. It is confirmed after the experiment that all the subjects have believed that the robot is moving and rotating automatically. The other two conditions are completely identical to those in the four conditions for the first hypothesis. Figure 7.1 summarizes all the conditions of the study.

7.5.5 Experimental Design

How can you judge a condition to be better than another? This judgment is difficult because reactions to a system vary across individual users. It frequently happens that the same system is liked by some but disliked by other users. So, you need to collect reactions from many users and see an overall tendency of reactions. Suppose you want to compare conditions A and B and have successfully collected ten people who will participate in the comparison as subjects. If you ask them to experience both conditions, you can obtain ten data for each condition. The average value of ten data shows the overall reaction to each condition. You can compare the average values of both conditions and judge which condition is better. This kind of comparison is called the "within-subjects experiment." The inherent problem of within-subjects experiment is that average values are not fully independent from each other, since they are produced by the same group of subjects. If a subject experiences condition A and then experiences condition B, the experience of condition A may affect data for condition B. An example of this is the order effect, which is caused by the repetition of the same task. If the subject does some task in condition A and then in condition B, the subject may be able to finish the task more efficiently in condition B than in condition A. Another example is the novelty effect, which is caused by an emphasis on the novelty of system. If the subject experiences a conventional user interface in condition A and experiences a state-of-the-art user interface in condition B, the subject is likely to become aware of the novelty of the user interface in condition B and rates it higher than the user interface in condition A. In a within-subjects experiment, it is difficult to deal with the novelty effect, but it is possible to prevent the order effect from affecting the results. A technique called "counterbalancing" is used for this purpose: you divide subjects into groups, each of which corresponds to one of the possible orders of conditions. The groups have to cover all the possible orders, and the number of subjects in each group should be as equal as possible. When you compare conditions A and B, half of the subjects experience condition A first, and the other half experience condition B first. When you compare three conditions, the number of possible orders is six. So, you need to prepare six groups of subjects, and they experience the three conditions in a different order.

The undesirable effects described above are caused by sharing the same subjects among conditions. The effects do not occur if you collect subjects separately for each condition and ask each subject to experience and rate only one of the conditions. If you do so, you can compare conditions without considering the independence of them, since data obtained in one condition is not affected by the other conditions at all. This kind of comparison is called the "between-subjects experiment," which enables relatively fair comparisons but has two problems. An obvious problem is that you need to collect a large number of subjects. When the number of conditions is N, a between-subjects experiment consumes N times as many subjects as a corresponding within-subjects experiment. If you test two factors each of which consists of two levels, the number of subjects necessary for the test increases fourfold in the between-subjects experiment, compared with a corresponding within-subject experiment. Another problem is less obvious but more serious: subjects have individual differences, e.g., in age, gender, knowledge, skills, and experiences. Of course, there are individual differences among subjects in a within-subjects experiment, too. However, that does not generate a problem, since the same subjects rate all conditions. Individual differences generate a problem in the between-subjects experiment, since a group of subjects who rate a condition may be very different from other groups who rate other conditions. For example, if one group includes a larger number of aged subjects than other groups, their poor eyesight and hearing ability may cause unnecessary biases in the evaluation of the system. If all the groups include almost the same number of aged subjects, this problem does not occur. So, in a between-subjects experiment, you should organize subjects with awareness of this problem. All groups of subjects should be equal in the age distribution, the ratio of males and females, and the percentage of people who have some special knowledge, skills, and experiences. This method of organizing subjects is called the "randomized block design." Unless you use this method, it is difficult for you to distinguish the effects of the relevant factors from the effects of individual differences. Even if you use the method, it is not easy to eliminate the effects of individual differences when subjects have various and large individual differences. Therefore, it is recommended that you collect people who are similar to each other and make the subjects as homogeneous as possible. This issue will be detailed in Sect. 7.5.7.

Telepresence Robots' Example To prevent the order effect and the novelty effect, a between-subjects design is employed. The first comparison is about two-by-two conditions in a between-subjects design, and the second comparison is about three conditions in a between-subjects design. Since the two comparisons share two conditions with each other, there are five conditions in total. So, the experiment requires five times as many subjects as the case in which comparisons are made in a within-subjects design. Figure 7.1 shows the situation where N subjects are assigned to each condition. 5 N subjects are necessary in total.

7.5.6 Task

The last subsection has discussed the difficulty of comparing conditions with regard to differences in subjects. This subsection discusses it with regard to task choice. A good task is one that can reveal the effects of the factors that lead to different results in different conditions, and a typical way of using a system is not necessarily a good task for it. A task that always leads to an extremely good or bad evaluation of the system should be avoided, since such a task conceals the effects of the factors. Suppose that you compare the attractiveness of two pet robots. If subjects are allowed to play with the robots for a long time, they may enjoy the robots too much and therefore rate them very high. If subjects are allowed to play with the robots only for a short time, they may enjoy the robots insufficiently and therefore rate them very low. In both cases, the factors that are different between the two robots do not lead to differences in evaluation. The former kind of phenomenon is called the "ceiling effect" and the latter kind of phenomenon the "floor effect." You should not choose a task that causes these effects. In the example being used here, you should appropriately control the length of time subjects take to play with the robots. Other kinds of tasks that should be avoided are related to the differences between laboratory study and field study. In a laboratory study, to obtain general findings, the task should be kept as simple as possible. But a simplistic task tends to be too artificial to reflect the real world. In a field study, to test a system naturally, the task should be the same as an actual manner of use in a real-world situation. If this is done, the task becomes too complex to obtain general findings. There is a trade-off between naturalness and generality, and the task should not be too artificial or too natural. To evaluate the pet robots in a realistic situation, for example, subjects should play with the robots at their own will, and experimenters should not instruct them to do anything. However, such a situation makes it difficult to observe the effects of the factors, since the ways of playing with the robots vary across subjects, and this variation is relevant to how they rate the robots. So, to reduce the variation, experimenters should instruct subjects to some extent. The ultimate method for reducing the variation is not by asking subjects to interact with the robots directly but by asking them just to see a footage in which someone plays with the robots. This kind of video-based evaluation is recently popular in human-robot interaction (HRI) research, but it is not generally recommended. This is because the observation of footage cannot provide the same experience as direct interaction with a robot. In a laboratory study, it is necessary to keep variations in the way of using a system as small as possible and to preserve the naturalness and reality of the experimental situation.

Telepresence Robots' Example The subjects remotely control the robot to travel to three tables and talk with an experimenter through a videoconferencing terminal that is mounted on the robot. At each table, the experimenter describes an object that is placed on the table. As shown in Fig. 7.2, the arrangement of tables is changed to generate four kinds of trajectories of robot's movement. In the move and rotate condition (as well as the automatic condition), the robot has to rotate and move in order to travel to the tables. The robot rotates to the right about 20° and moves forward about 1 m to approach the first table. To approach the second and third tables, the robot rotates to the right about 45° and moves forward 1 m. In the move condition where the tables form a line, the robot only needs to move. The subjects move the robot forward about 1 m to approach the presenter. In the rotate condition where the tables form a circle, the robot only has to rotate. The subjects rotate the robot to the right about 20° in order to aim at the first table and to the right about 60° in order to aim at the second and third tables. In the fixed condition, the subject talks with the experimenter at the same table three times and does not need to control the robot.

7.5.7 Subjects

What kind of person is suitable as a subject in the evaluation of a system? The answer heavily depends on what kind of study is relevant to the evaluation; there is no perfect guideline. Experimenters are sometimes unwilling to screen subjects since they want to increase the number of subjects. As a matter of fact, however, the results of an experiment are affected by particular subjects who participate in it. This subsection discusses the suitability of people as subjects. As mentioned in Sect. 7.5.5, it is recommended to collect subjects who are similar to each other and keep subjects as homogeneous as possible. In many cases, undergraduate students are used as subjects. One reason for this is that it is easy to use them as subjects in experiments at universities, but this is not the only reason. Another reason is that they are neither too old nor too young. Aging involves a decrease in perceptual abilities, e.g., eyesight and hearing ability, and inevitably affects the evaluation of the user interface of a system. If subjects are too young, they may not be able to understand how to use a system or follow experimenters' instructions. Undergraduate students are mostly immune from these problems. Another reason for using undergraduate students is that they are educated but do not have professional knowledge. Teachers and graduate students have some professional knowledge, which enlarges individual differences and distorts the evaluation of system. For example, people who are involved in the research or development of robotic systems may react to pet robots differently from others. Unless it is specifically required to collect reactions from various kinds of people, undergraduate students are a good choice.

Telepresence Robots' Example Thirty-five undergraduate students have participated in the experiment. Their ages range from 18 to 24 years. The experiment is a between-subjects design with five conditions, and seven students participate in each condition, as shown in Fig. 7.2. Three female and four male students participate in the move, rotate, and fixed conditions. Two female and five male students participate in the move and rotate and automatic conditions.

7.5.8 Data Collection

Data collection means measuring the subjects' reactions to a system. There are two kinds of measurements, and they are very different. One is objective measurement in which experimenters observe the behaviors of subjects by using various means of sensing and recording (e.g., data logging, video recording, and motion tracking). The other is subjective measurement in which experimenters ask subjects about what they think and feel while they are using a system (e.g., questionnaires and interviews). It is widely assumed that objective measurement is more scientific and thus more recommendable than subjective measurement. However, the reality is sometimes the opposite. The results obtained from objective measurement become subjective when researchers interpret the meaning of collected data. In objective measurement, collected data are observed behaviors of subjects, and they are usually not self-explanatory and require some interpretation. The larger the gap between the hypotheses and the collected data is, the more the results become subjective. How do you measure the attractiveness of a pet robot objectively? Do you measure playtime with it, the frequency of smiling, or the average distance between the subject and the robot? You must subjectively interpret a long playtime, high frequency of smiling, and short average distance as indications of attractiveness. The problem here is that the same behaviors (playing enthusiastically, smiling, and coming close) can be exhibited for multiple reasons, i.e., attractiveness is not the only reason. So, subjective interpretation can easily lead to misinterpretation and wrong results.

The results obtained from subjective measurement can be more objective and scientific than those obtained from objective measurement, since collected data in subjective measurement are subjects' answers to experimenters' questions; they are self-explanatory and require little interpretation. However, data are unreliable unless the following two conditions are satisfied. First, subjects need to correctly understand what experimenters intend to ask. When they misunderstand questions, their answers are skewed and produce meaningless results. Second, subjects need to be unaware of what kinds of results are to be expected. When subjects are aware of them, it is difficult for subjects to answer straightforwardly. Subjects may try to meet the expectations or do the opposite if they are perverse. Another point to which attention should be paid in subjective measurement is the trade-off between the ease of quantitative analysis and the richness of content. When the study employs questionnaires or structured interviews as a subjective measurement, it is easy to analyze collected data quantitatively, since experimenters always ask the same questions for all subjects. By contrast, unstructured interviews, in which experimenters choose or create a question to ask according to a subject's answers, help to understand the subject's reaction to the system more deeply than questionnaires or structured interviews do. However, such a deep understanding tends to be specific to an individual subject, and it is difficult to analyze data quantitatively and retrieve general findings from them.

Since objective measurement has an inherent limitation, subjective measurement is often the only choice. Obviously, objective measurement is useless for measuring what subjects think and feel if it is not manifested by observable behaviors, e.g., facial expressions, gestures, and other nonverbal behaviors. It is still very hard to use brain activity measuring systems for this purpose. These systems are costly and difficult to use (but these problems will be solved in the near future). The large gap between measured brain activities and psychological phenomena is a serious problem, since the gap generates the interpretation problem discussed above. So, subjective measurement will surely survive for a long time as a valuable tool for data collection.

Telepresence Robots' Example A subjective measure is used to measure the degree of presence of the presenter who has described the objects on the tables. As shown in Fig. 7.3, the questionnaire asks the extent to which the following statements match the impressions the subject has about the task: "I felt as if I were talking with the presenter in the same room," and "I felt as if I were viewing the presenter in the same room." These questions are rated on a 9-point Likert scale where 1 = strongly disagree, 3 = disagree, 5 = neutral, 7 = agree, and 9 = strongly agree.

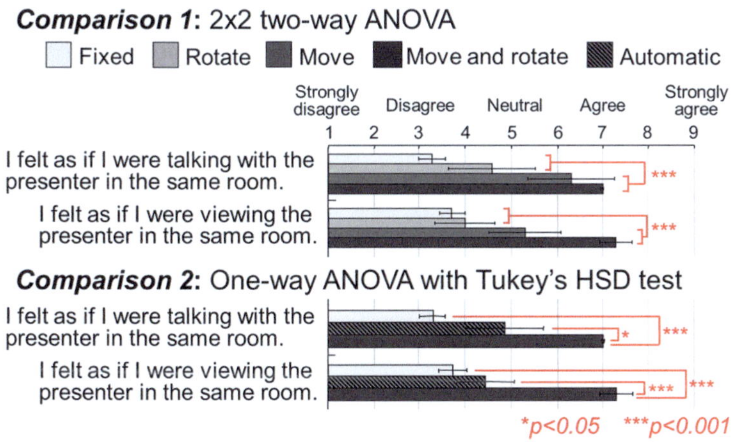

Fig. 7.3 Data collection and analysis

7.5.9 Analysis

Statistical techniques are usually used for revealing the differences between conditions. In general, you calculate the average and variance of scores for each condition and then judge that there is a significant difference between the conditions when their average scores largely differ from each other and simultaneously their score variances are small. Suppose that you ask subjects about the degree of attractiveness of cat and dog robots. If you find the average score of the cat robot is much higher than that of the dog robot and both scores are associated with small variances, then you can conclude that the cat robot is more attractive than the dog robot. This is a simple statistical technique, and there are more complex ones. However, complex techniques are double-edged swords, since they can reveal the subtle differences between conditions but blur the relationships between collected data and results. Complex techniques may give rise to a suspicion that researchers have tried many ways of processing data and then arbitrarily employed one of them as the analysis method because it generates the results that the researchers prefer. Thus, complex statistical techniques are a powerful tool but not necessarily a good choice. It is recommended to keep the method of processing data as simple as possible.

Usually, the quantitative analysis that uses statistical techniques to process numerical data is regarded as a rigorous method of analysis. However, it may easily produce wrong results if researchers ignore what numerical data actually mean. Unfortunately, such ignorance is observed frequently even in high-quality research papers. This problem is similar to the interpretation problem. Anecdotal evidence that is obtained from interviews and open-ended responses to questionnaires are usually regarded as subsidiary results, but these literal data are very helpful for avoiding misunderstanding the meanings of numerical data. In the pet robots' example, the robot's appearance may not be the cause of the difference in attractiveness. Another factor, e.g., size, may be the cause. Literal data are useful for detecting such a hidden factor. Furthermore, the robot itself may not be the cause, and the imperfection in controlling the experimental environment may be the cause. For example, the robot may be rated low if the room where subjects play with it is very uncomfortable. Literal data also help to become aware of this kind of imperfection.

Telepresence Robots' Example Seven subjects participate in each condition, and so seven data sets are obtained for each condition. To examine the first hypothesis, the move and rotate, move, rotate, and fixed conditions are compared by 2×2 two-way between-subjects ANOVA. To examine the second hypothesis, the move and rotate, the automatic, and the fixed conditions are compared by one-way between-subjects ANOVA with Tukey's HSD test. The results of these comparisons are shown in Fig. 7.3, in which each box represents the mean value of the responses to each statement in each condition and each bar represents the standard error of the mean value. The first comparison reveals strong main effects of the forward-backward movement factor, no significant main effect of the right-left rotation factor, and no significant interaction between these factors. The movement significantly strengthens the feeling as of talking with the presenter in the same room ($F(1,24) = 15.753$, $p < 0.001$) and viewing the presenter in the same room ($F(1,24) = 18.951$, $p < 0.001$). These results support the first hypothesis. In the second comparison, a significant difference is found in the feeling as of talking with the presenter in the same room ($F(2,18) = 13.566$, $p < 0.001$). Multiple comparisons show that the feeling is significantly stronger in the move and rotate condition than in the automatic condition ($p < 0.05$) and the fixed condition ($p < 0.001$), but there is no significant difference between the automatic and fixed conditions. A significant difference is found in the feeling as of viewing the presenter in the same room ($F(2,18) = 18.314$, $p < 0.001$). Multiple comparisons show that the feeling is significantly stronger in the move and rotate condition than in the automatic condition ($p < 0.001$) and fixed condition ($p < 0.001$) but again that there is no significant difference between the automatic and fixed conditions. These results are consistent with the second hypothesis.

Exercises

1. This chapter has introduced and described an example of telepresence robot in a certain way. In the same way, summarize the goal, hypotheses, factors, conditions, experimental design, task, subjects, data collection, and analysis of some other study on robotic telepresence technology (see Nakanishi et al. 2009, 2011, 2014).
2. Plan your own study that evaluates the user interface of some interactive system, and describe its goal, hypotheses, factors, conditions, experimental design, task, subjects, data collection, and analysis.

References

Ishiguro, H., Trivedi, M.: Integrating a perceptual information infrastructure with robotic avatars: a framework for tele-existence. In: Proceedings of the International Conference on Intelligent Robots and Systems (IROS 99), pp. 1032–1038 (1999)

Jouppi, N.P.: First steps towards mutually-immersive mobile telepresence. In: Proceedings of the International Conference on Computer Supported Cooperative Work (CSCW 2002), pp. 354–363 (2002)

Nakanishi, H., Murakami, Y., Kato, K.: Movable cameras enhance social telepresence in media spaces. In: Proceedings of the International Conference on Human Factors in Computing Systems (CHI 2009), pp. 433–442 (2009)

Nakanishi, H., Kato, K., Ishiguro, H.: Zoom cameras and movable displays enhance social telepresence. In: Proceedings of the International Conference on Human Factors in Computing Systems (CHI 2011), pp. 63–72 (2011)

Nakanishi, H., Tanaka, K., Wada, Y.: Remote handshaking: touch enhances video-mediated social telepresence. In: Proceedings of the International Conference on Human Factors in Computing Systems (CHI 2014), pp. 2143–2152 (2014)

Nakanishi, H., et al.: Minimum movement matters: impact of robot-mounted cameras on social telepresence. In: Proceedings of the International Conference on Computer Supported Cooperative Work (CSCW 2008), pp. 303–312 (2008)

Paulos, E., Canny, J.: Social tele-embodiment: understanding presence. Auton Robot **11**(1), 87–95 (2001)

Prussog, A., Muhlbach, L., Bocker, M.: Telepresence in videocommunications. Annual Meeting of Human Factors and Ergonomics Society, pp. 180–184 (1994)

Vogeley, K., Fink, G.R.: Neural correlates of the first-person-perspective. Trends Cogn. Sci. **7**(1), 38–42 (2003)

Chapter 8
Robotics for Safety and Security

Tatsuo Arai and Hiroko Kamide

Abstract Certain societal issues, such as the rapidly growing aged society and the increase in crimes, imply that service robots are required to operate in a safe manner. It is important to discuss the two aspects of the safety of robots: physical safety and mental safety. A kind of physical safety is discussed by showing an example of technology: a humanoid robot capable of carrying a wheelchair user. As for mental safety aspect, a new psychological scale is developed to quantify general impressions toward humanoids, using 11 different humanoid robots and 3543 Japanese respondents. It is revealed that nine factors are used for evaluating the general impressions of robots: familiarity, repulsion, performance, utility, motion, sound, voice, humanness, and agency.

Keywords Safety and security • Service robot • Mental safety • Pushing manipulation by a humanoid • Human factors

8.1 Social Demands for the Robot

There are many societal issues in the current Japanese society. The first most serious issue is the population problem that arises from Japan's rapidly growing aged society. According to the data (Population statistics, National Institute of Population and Social Security Research) shown in Figs. 8.1 and 8.2, there will be too many senior people and too few younger generations in the near future. This problem causes other issues, for example, lack of labor power, a boost in medical expenses, etc. For this reason, we need to support people's daily life and encourage women and even seniors to actively engage in the society, in order to improve the nation's productivity. People have a variety of requirements for household and social life, such as good nursery support services for babies and medical care services for

T. Arai (✉)
Graduate School of Engineering Science, Osaka University, Toyonaka, Osaka, Japan
e-mail: arai@sys.es.osaka-u.ac.jp

H. Kamide
Research Institute of Electrical Communication, Tohoku University, Sendai, Miyagi, Japan

M. Kasaki et al. (eds.), *Cognitive Neuroscience Robotics A*,
DOI 10.1007/978-4-431-54595-8_8

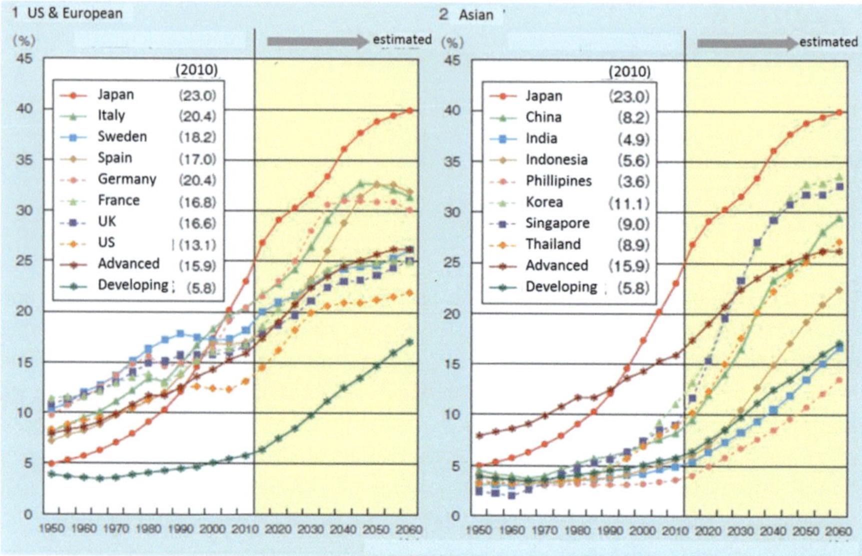

Fig. 8.1 Percentage of elderly population by country

Fig. 8.2 Estimated Japanese population

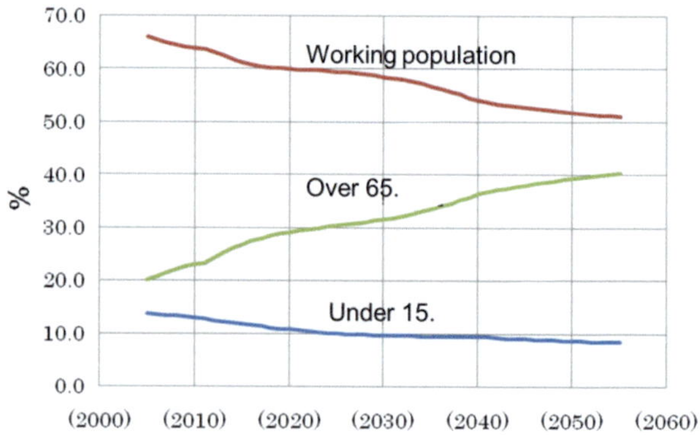

seniors. Although these services should be provided by humans, new technology will be developed to cover the serious lack of labor power.

As far as security is concerned, we have been confronting the worst situation. Although Japan is generally believed to be safer than other countries, this has turned out to be an illusion. For these twenty years, the number of crimes in Japan has been increasing drastically, and the increase in vicious crimes has brought great anxiety to the Japanese society (The White Paper on Police 2010). Japan was once proud of being the securest country in the world, but it is now suffering crimes that exist typically in advanced countries due to the separation between the rich and the poor brought about by borderless internationalization and globalization of economy.

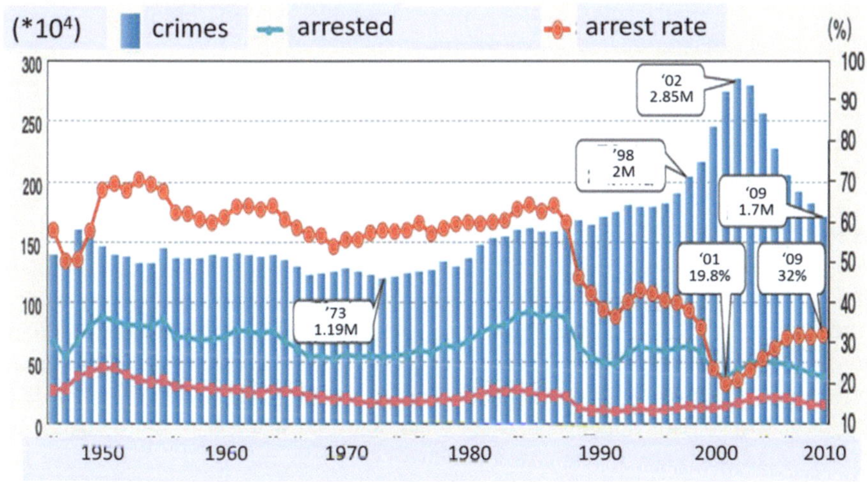

Fig. 8.3 Number of crimes and arrest rate in Japan

Many monitoring cameras have been set everywhere in big cities for the purposes of preventing these crimes and identifying criminals quickly. Although the security system with cameras can provide 24-h monitoring, people feel that they are always watching them, and even worse, it may cause some sort of mental stress. It must be required that machines and mechanical systems protect and support our social and private activities securely but amicably. If any robot were to work in a house and/or in public space, we would expect it to provide us with safe and secure life by using its amicable and dependable functions (Fig. 8.3).

It is important to consider how *robotics technology* (RT) should be used in the future society and in particular to consider questions such as how RT should support human life and work, how it should protect people, and how it should keep the social security. As was suggested in the METI's Technological Strategy Road Map, useful service robots are required to support us in friendly manners, and the development of such robots involves addressing current and future issues, such as "increase of seniors in aged society;" "decrease of labor powers;" "safe, secure, and dependable society;" and "realization of comfortable and high-quality life."

While service robots will be utilized in our society, some people show distrust in them. In Europe and the US, people strongly oppose to the introduction of robots into their private life. There are lots of pros and cons for using robots for nursing, caring, and welfare tasks, and we, engineers, always encounter questions, for example, "why do we need machines for human tasks," "how we can reserve the human sovereignty," etc. Every country and region has its specific background, history, culture, and religion. Thus, we need to be careful when we propose applications of service robots. It is important to discuss the safety of robots from the viewpoints of not only technology but also humanity and social science. All of these viewpoints should be taken into account in order to achieve the commercial success of the service robot industry. In what follows, we will discuss two aspects of the robot safety issues: *physical safety* and *mental safety*.

8.2 Safeties Required for Service Robot

A service robot is defined, for example, in the World Robotics (2007), as "a robot which operates semi- or fully autonomously to perform services useful to the well-being of humans and equipment, excluding manufacturing operations." There are many application areas of service robots, such as housework support, medical care, nursing, rehabilitation, reception, security, cleaning, construction, repairing, agriculture, and rescue. Here we consider only limited areas where robots interact closely with humans. In such areas, the physical safety of robots is the most concerned issue. In the case of industrial robots, the labor safety codes are established by the Japanese Ministry of Health, Labour and Welfare. They are effective to guarantee that no robot hurts or injures users, simply by requesting building fences around industrial robots to keep humans away from them. But this maneuver does not work in the case of service robots, since they, by their nature, work quite close to and/or in contact with users. Safety engineering should be studied and potential risks should be assessed carefully. Service robots should not go out of control, should not hurt and injure user even when they collide, and should have foolproof and fail-safe systems.

The Japanese Ministry of Economy, Trade and Industry encourages to develop commercial service robots and expects there to be a good market for them in the future. The Ministry has been supporting the development of service robots by establishing the safety codes for service robots and funding the R&D through a promotion of the NEDO Project for Practical Application of Service Robots. The Ministry has founded the Robot Safety Center, and it is responsible for both the verification of the safety of every service robot product and its official authorization. On the other hand, the safety codes are still under discussion. The ISO13482 will be issued soon, and it will cover hazard identification and risk assessment, safety requirements, protection measures, and information on the usage of "personal care robots." On the basis of the draft of ISO13482, the Japan Quality Assurance Organization has approved the safety of the robot suit "HAL" and issued its assurance in February 2013. This is the first case of an assured service robot.

Even if a service robot can have necessary physical safety means, users may not always feel comfortable with it. Users themselves must understand that the robot will not hurt them; otherwise, they will be scared and anxious of the robot. Thus, the "mental safety" of robots, as we call it, is another important issue which we should be concerned with in considering the service application of robots. Figure 8.4 summarizes the important considerations in ensuring the safety of industrial and service robots.

Fig. 8.4 Comparison between two robots

8.3 An Example of Supportive Robot: Humanoid Pushing a Wheelchair

We will discuss technical issues concerning the practical application of humanoid robots by using the wheelchair assistant robot as an example. The HRP-2 is a humanoid robot, and it is able to do the task of pushing a wheelchair with a person in it; for this task, the robot needs to exert enough force to push the wheelchair and to keep it moving stably. In the following sub-sessions, we will show the robust and simple control strategies required to push heavy objects, such as a wheelchair with a person in it.

8.3.1 Target Object

Pushing manipulation by a humanoid robot has been studied. In particular, balancing control has been studied with the assumption that the pushing force is external to preplanned action. One of us proposed a method for pushing an object in dynamic walking (Takubo et al. 2005). The main focus of this study was on walking patterns and was not directly concerned with pushing force generation.

When a human pushes an object, he/she recognizes the object and poses his/her whole body for generating strong pushing force. Generally, a tool with wheels, such as a cart, is often used to transport objects, and we can easily know how to move the tool when we use it; most prominently, we gain information on how to move it and accordingly control the force we exert on it. By the same token, humanoid robots should have prior information as to how to change its pushing action in accordance with the required pushing force. Here we propose a model for pushing manipulation that focuses on pushing force rather than walking patterns. On the other hand, how much force is required to move a given object is usually unknown, and the force generating algorithm may be complicated if it takes account of robot's balance and required pushing force. In order to achieve a real-time control of humanoid robots for its safe application, we need to avoid the complication of algorithm and should instead aim at a simple control scheme.

When we figure out the condition of the target object, we acquire certain information on that object. The question is what information it is and how to define it. When humans recognize the shape of an object, they usually know the grasping points of the object and then exert force on them. For simplicity, we assume that the target object is simple, and that robots can recognize grasping points definitely. However, we do not assume that they know how much force is required to move the object. In more detail, we distinguish between the information robots acquire and the information they do not, as follows:
Acquired information:

1. Contact or grasping points of the object
2. Limiting force that keeps the object stationary even if some force is put on the contact point

Unknown information:

3. Mass of the object
4. Force required to move the object

These types of information are general, and hence we assume that any pushing task requires processing them. To advance the discussion further, we make another assumption that manipulated objects have stable points whose characteristics do not change if a robot gives any load (e.g., its weight) on them.

8.3.2 Basic Principle

The stability criteria that use the multipoint region supported by hands and legs are investigated in the previous studies (Takenaka et al. 1998; Harada et al. 2004). By using the methods these studies have developed, a humanoid robot is capable of reaching its hand further and standing up from, for example, a chair. We will consider the stability of pushing manipulation by using multipoint contact states. First, we assume that when a robot leans on a fairly large object, it can be supported by both its hands and legs.

Figures 8.5 and 8.6 show the top, side, and rear views of contact states. For basic stability condition 3, second, we focus on the two regions in the x-y plane which are the projection of the supporting area composed of both hands and one of the legs.

When the center of mass (CoM) position is located in these regions, the robot does not fall down even if the right or left foot takes off the ground. If it is possible to keep the above CoM condition, a humanoid robot does not need to move the CoM position to walk; it can walk by using simple foot trajectories.

We define the stability criterion in terms of the CoM position and multipoint supported regions. We consider stability by projecting the CoM position in the x-y

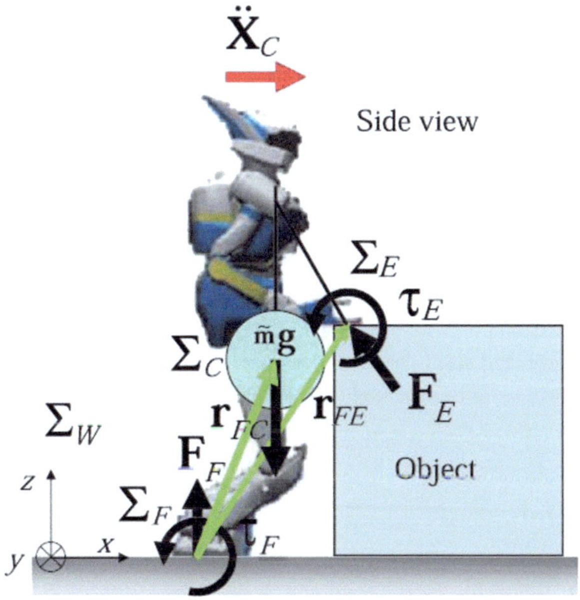

Fig. 8.5 Single-mass model and contact states (*side view*)

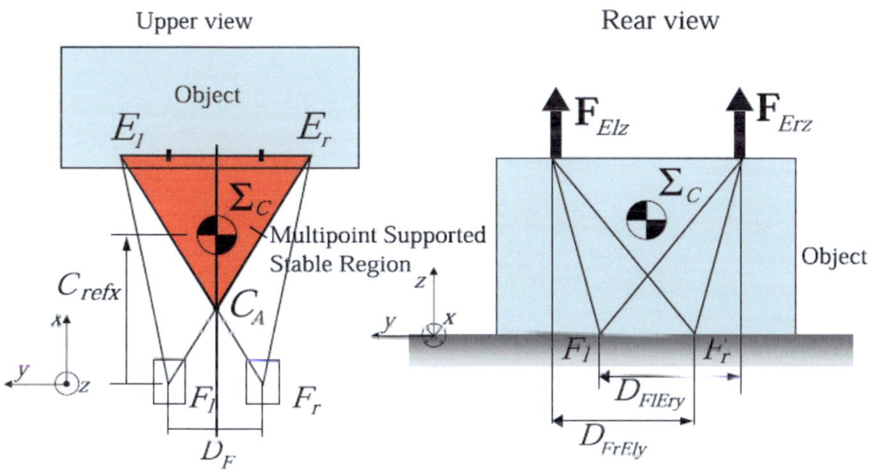

Fig. 8.6 Stable region and contact states (*upper* and *rear views*)

plane to the upper view, as is illustrated in Fig. 8.3. E_r and E_l are the tip positions of the right and left manipulators, respectively; and F_r and F_l are the right and left feet positions, respectively. The intersection of lines E_rF_l and E_lF_r is C_A. The overlapped delta region between $E_rE_lF_r$ and $E_rE_lF_l$ is defined as the multipoint supported stable region (MSSR). Assuming that the static balance and the CoM position locate in the MSSR, the humanoid robot can be supported by either right or left foot without falling down. To calculate the stable CoM position in the MSSR, we define the target CoM position C_{refx} as follows:

$$C_{refx} = \frac{D_F}{D_F + D_E} K_C \, \mathbf{r}_{EFx} \tag{8.1}$$

D_F is the length between the two feet in the y direction. D_E is the length between the two hands in the y direction. $\mathbf{r}_{EFx}$ is the manipulator tip position in the x direction of the foot coordinates. K_C is a coefficient to define the CoM position. When $K_C = 1$, the CoM position locates on the intersection point C_A. When $K_C > 1$, the CoM position locates in the multipoint supported stable region. If we regard the multipoint supported stable region as a support polygon, we can judge about the stability of the robot by using the stability margin.

8.3.3 Stability and Pushing Force

In this section, we consider a necessary requirement for balance of a humanoid robot. If the target zero moment point (ZMP) is given in the support polygon consisting of two legs and an external force acts on the two manipulator tips, then dynamical complement zero moment point (DCZMP) is calculated as follows:

$$\begin{aligned} r_{FCx} = \Big(& \left(\tilde{m}g_z - F_{Ez} + \dot{P}_z\right) ZMP_{xd} - \dot{L}_y + r_{FCz}\dot{P}_x \\ & + \tau_{Ey} + \tilde{m}r_{FCz}\ddot{X}_{Cx} - r_{FEx}F_{Ez} - r_{FEz}F_{Ex}\Big) / \tilde{m}\left(g_z - \ddot{X}_{Cz}\right) \end{aligned} \tag{8.2}$$

$$\begin{aligned} r_{FCy} = \Big(& \left(\tilde{m}g_z - F_{Ez} + \dot{P}_z\right) ZMP_{yd} - \dot{L}_x + r_{FCz}\dot{P}_y \\ & - \tau_{Ex} + \tilde{m}r_{FCz}\ddot{X}_{Cy} - r_{FEy}F_{Ez} - r_{FEz}F_{Ey}\Big) / \tilde{m}\left(g_z - \ddot{X}_{Cz}\right) \end{aligned} \tag{8.3}$$

where P and L denote the whole-body momentum and angular momentum at the foot frame Σ_F, respectively. $\mathbf{r}_{FC}$ is a vector from the origin of the foot coordinate system to the CoM position, and $\mathbf{r}_{FE}$ is a vector from the origin of the foot coordinate system to the manipulator tip position. The subscripts x, y, and z denote directions in the world frame. Equations (8.2) and (8.3) show the projection of the CoM position in the x-y plane, and we define them as the DCZMP. It consists of three terms: the first is calculated by the target ZMP, the second is calculated by the force acting on the two hands, and the third is calculated by the CoM acceleration. The target ZMP trajectory and CoM acceleration are defined by the feet trajectories. In addition, the maximum pushing force, FEmax, is limited by the capacity of the robot actuator and the mass or weight of the target object. Thus, the tolerable area of the CoM position with regard to the current motion is obtained. Generally, the tolerable area becomes larger as the limiting force becomes larger. By locating the CoM position in the tolerable area, the humanoid robot does not fall down. Let us assume that the

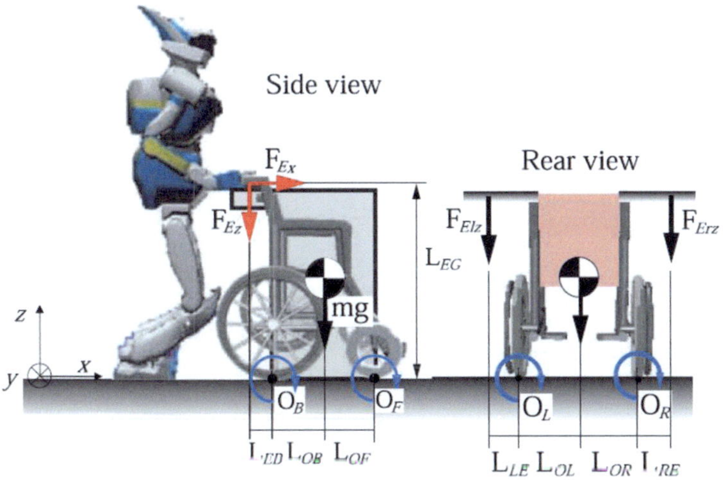

Fig. 8.7 Model of a target object

friction forces between an object and hands are large or the robot holds an object firmly, and that the acceleration in the z direction is zero. Then, the pushing force in the forward direction, F_{Ex}, is expressed as follows:

$$\tilde{m}gC_{refx} - \tilde{m}\ddot{X}_{Cx}r_{FCz} - F_{Ez}r_{FEx} - F_{Ex}r_{FEz} = 0 \tag{8.4}$$

$$F_{Ex} = \frac{\tilde{m}gC_{refx} - \tilde{m}\ddot{X}_{Cx}r_{FCz} - F_{Ez}r_{FEx}}{r_{FEz}} \tag{8.5}$$

F_{Ex} is the maximum value when the CoM has acceleration ($\ddot{X}_{Cx}$) in the x direction.

Next, let us introduce the condition of a target object that keeps the three-point support consisting of two hands and a leg. We assume that the object does not fall down even if the robot leans on it with both hands. The imaginary ZMP (Vukovratovic et al. 2001) of the object must be in the support area of the object for it to move stably, but it is difficult to know the real ZMP of the object. Here, we assume a simple model that the object with the mass m is supported in a rectangular area, as shown in Fig. 8.7. Let F_{Ez} be the hand reflect force in the vertical direction. Assuming a moment equilibrium equation about O_B, the stability of the object is defined as follows:

$$mgL_{OB} - F_{Ez}L_{EB} > 0 \tag{8.6}$$

where L_{OB} denotes the length from the CoM position to the rear end of the support area and L_{EB} is the length from a contact point of a hand to the rear end of the support area. Given an equation of the moment about O_F in the x-y plane, the pushing force F_{Ex} in the forward direction is expressed as follows:

$$F_{Ex}L_{EG} - mgL_{OF} < 0 \tag{8.7}$$

By the same token, the conditions for the right and left stabilities are determined as follows:

$$F_{Erz}L_{RE} - F_{Elz}\left(L_{LE} + L_{OL} + L_{OR}\right) - mgL_{OR} < 0 \tag{8.8}$$

$$F_{Erz}\left(L_{RE} + L_{OR} + L_{OL}\right) - F_{Elz}L_{LE} + mgL_{OL} > 0 \tag{8.9}$$

We assume that objects meet these conditions in simulations and experiments.

8.3.4 Walking Pattern

In the case of multipoint support, a humanoid robot can use the same walking strategy as a multi-legged robot does. Figure 8.8 shows a foot trajectory of a humanoid robot. The swing leg follows an elliptic orbit and the support leg moves in the horizontal direction. Both feet remain parallel with the ground. Let l_s, θ_s, and x_s be a step length, the angle for representing a position on the elliptic orbit, and the position of the support leg in the swing phase, respectively. We define the relation among θ_s, r_s, and x_s as follows:

$$\theta_s = \pi\left(1 - x_s/l_s\right) \tag{8.10}$$

$$r_s = r_s\left(\theta_s\right) \tag{8.11}$$

Fig. 8.8 Foot trajectory in walking

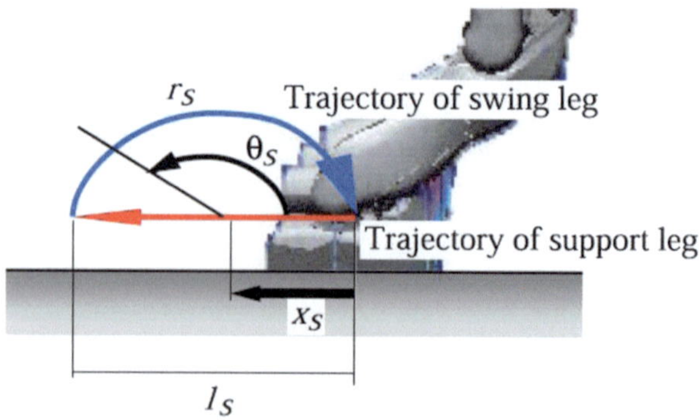

r_s is the position of the swing leg. The humanoid robot moves following the motion of the support foot, and this foot motion can be changed freely. We can obtain the motion of the walking for pushing, although the pushing force is limited as in Eq. (8.5). Thus, we will apply an impedance control to the CoM position control for modifications of the travel position according to the hand reflect forces. Given Eq. (8.5), the acceleration of the CoM is given in the following equation:

$$\ddot{X}_{Cx} = \frac{\tilde{m}gC_{refx} - F_{Ez}r_{FEx} - F_{Ex}r_{FEz}}{\tilde{m}r_{FCz}} \tag{8.12}$$

Let us assume that the acceleration of the supported foot in the body frame is the same as the acceleration of the CoM in the world frame. If the acceleration calculated by Eq. (8.12) is smaller than the supported foot acceleration, we limit the acceleration of the foot as the value of Eq. (8.12). This is simply for impedance control of the CoM. We can easily use this control to change the acceleration of travel without considering complicated balance control of the humanoid, biped robot. When the external force changes, the humanoid robot can change its pushing force by the motion of its whole body with the impedance control of the CoM. If the reflection force is too large to push the object, the humanoid robot can step back according to the measured force. The control of the CoM and whole body is achieved by the resolved momentum control (Kajita et al. 2003).

8.3.5 Experiments on HRP-2

We applied the proposed method to HRP-2. We selected a 3 kg paper box fixed on a 12 kg table as a task object. The object can be moved in the horizontal direction by a 20–30 N force. In this experiment, we examined whether the proposed method was able to do a pushing motion in accordance with the required force. We set a 10-step walking pattern and blocked the object from moving in the forward direction during the robot's walking. Figures 8.9, 8.10, and 8.11 show snap shots of the experiment, the pushing force, and the foot trajectory in the x direction in the body coordinate frame. The humanoid robot did not slip. In Fig. 8.11, the humanoid robot was not able to step forward for 12–15 s when the object was blocked. However, we found that the robot was able to continue to push the target after the block was removed. The robot was able to adapt to the required force by exerting the impedance control of the CoM position in real time. The robustness of the proposed method was confirmed.

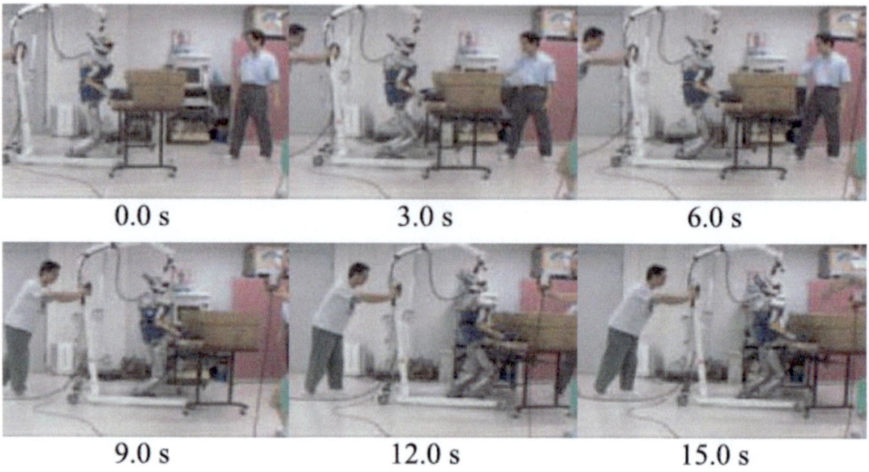

Fig. 8.9 Experimental results (object mass, 18 kg)

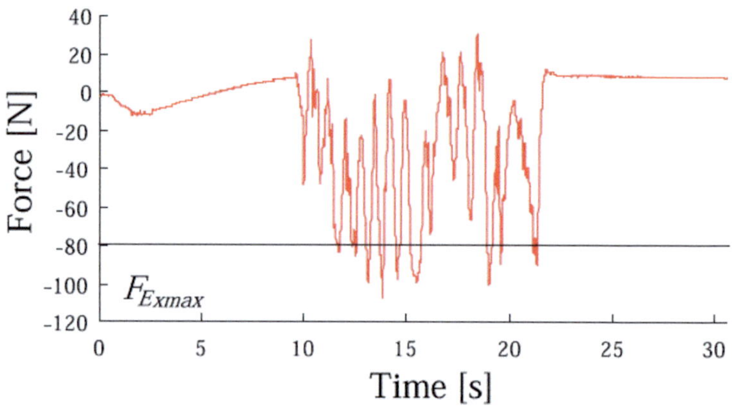

Fig. 8.10 Experimental results of pushing force in the walking direction

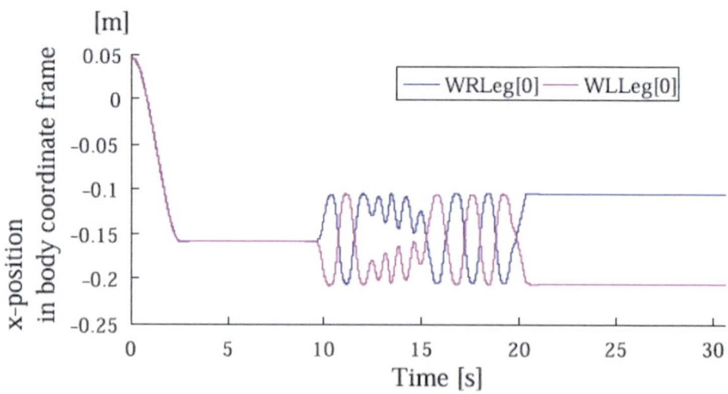

Fig. 8.11 Experimental result of foot trajectories in body coordinate frame

8.4 Development of a Psychological Scale for General Impressions of Humanoid

8.4.1 Methods for Evaluating Psychological Impressions About Humanoids

Humanoid robots are expected to companionably coexist with humans in the near future. While it is important to develop technologies to improve robot systems by engineering approach, human psychology has significant implications for the realization of humanoid robots that can companionably exist in the neighborhood of humans. Hence, we should take into account both the development of engineering technology and inquiry into psychological effects.

There are methodological limitations in the evaluation of psychological impressions of robots. For example, some studies used psychological scales derived from the semantic differential (SD) method. The SD method is originally developed for the evaluation of perceptions among humans (Inoue et al. 2007; Negi et al. 2008). It uses adjectives selected to measure the human's personality or impressions. Kamide et al. (2013) have found that people use different perspectives for evaluating a humanoid robot and developed a particular psychological scale for a type of robot. This is just a first step. In order to evaluate subjective impressions of humanoid robots more generally, it is necessary to clarify basic general perspectives on humanoid robots and develop a new scale for evaluating general impressions about them.

It is noteworthy that Bartneck et al. (2009) develop a psychological scale for the evaluation of robots and suggest that there are five aspects of robots that should be evaluated. These aspects are chosen because they are often used in engineering or human-robot interaction study. As far as the psychological acceptability of robots is concerned, ordinary people's perspectives on robots are also important, for it is ordinary people who are potential users of robots. Therefore, we focus on ordinary people's opinions about humanoid robots and aim to develop a psychological scale for robots in general on the basis of the psychological scale that is originally made for the evaluation of humanoid robots in particular.

We will report our two studies in what follows. In Study 1, we focus on ordinary people's general perspectives on humanoid robots and clarify the relative general standards for evaluating them. Then, in Study 2, we develop a new scale for evaluating general impressions of humanoid robots on the basis of the standards we derive from Study 1.

To achieve the goals of these studies, impressions about humanoid robots must be collected from a wide swath of the population; this is necessary to avoid the limitation that the obtained data do not reflect general opinions. To ensure that our data is highly representative of the population, we collect data from Japanese males and females in various age groups, ranging from the teens to the 70s. We conduct a survey in three areas of Japan (Tokyo, Osaka, and Fukuoka). Multiple types of

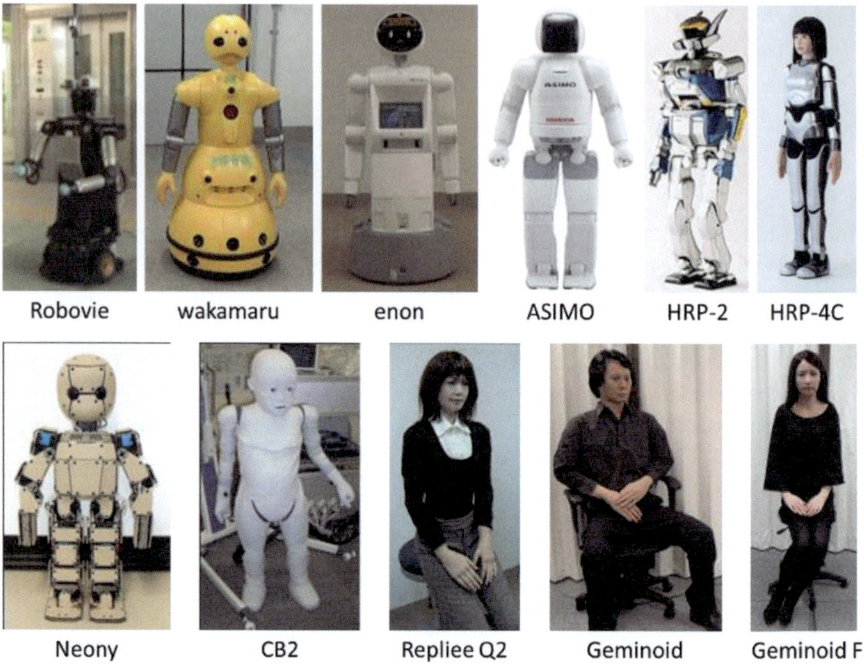

Fig. 8.12 Pictures of robots used in this study; Neony was formally known as M3-Neony and "CB" refers to the child robot with biomimetic body

humanoid robots should be used to establish general standards. For this reason, we use 11 different humanoid robots, including wheeled-walking robots, biped-walking robots, and androids.

8.4.2 *Study 1: Categorization of Natural Impressions to Discover the General Standards for Evaluation*

In Study 1, we aimed to collect qualitative data of ordinary people's thoughts or feelings about humanoid robots to discover the basic standards. Figure 8.12 shows the 11 robots presented as stimuli in this study. We presented movies of the robots to participants. The movies contain the scenes in which each robot functions in its favorable environment, and so participants can form opinions about it and its functions.

We divided the 11 robots into four groups and conducted a survey on each group. Each survey involved 225 participants, and the number of total participants was 900. A facilitator instructed the participants to watch a movie and then describe their impressions in distributed open-ended questionnaires. A questionnaire asked them to describe in detail what they thought or felt about a humanoid robot: "How do you

Table 8.1 Explanation of obtained categories

Category	Contents of descriptions
Familiarity	
Familiarity	General preferences for robots, motions, and designs
Repulsion	Anxiety or a sense of aversion that humans might be replaced by robots with regard to work or existence
Technology	
Utility	Clear goals of robots' tasks, usefulness, and costs
Performance	General evaluations of performances including interactions, sensations, and intelligence
Motion	Smooth, natural, and dynamic motions or stability and balance of motions
Sound	Soft sound of motors and clear voice
Humanness	
Humanness	General humanness including designs, motions, and voice
Emotion	Expression of feeling by face and eye movements or lip-synching of words; perceived emotion or will of robots
Robotness	General appearance of robots including designs, motions, and voice
Others	Direct descriptions of the contents of the movies or robots, controllability, vulnerability, experiences with robots, views about Japanese technology, etc.

Note that all categories except repulsion contain both positive and negative values

feel or what do you think about this robot? Describe your impressions as concretely as possible. If you have multiple impressions, describe each in one sentence."

8.4.2.1 Categories Obtained from Natural Impressions

We obtained a total of 17,776 descriptions. To analyze the qualitative data, it is important to be objective about categorization. We had five psychologists for analysis of the data, and they agreed on the way of categorization used below. We had ten categories after categorization, and our further discussion led us to group these categories into three major groups: familiarity, technology, and humanness (Table 8.1). To develop an appropriate psychological scale, we checked what the participants wrote on each of the three categories and classified them into 50 items of a closed-ended questionnaire (each category has 4–7 items). In Study 2, this scale is refined by statistical analysis.

8.4.3 Study 2: Factor Analysis to Clarify the Factor Structure and Reliability of the Scale

In this study, we aimed to analyze the new scale we derived from Study 1 by objective statistical methods. We asked participants to evaluate the same movies

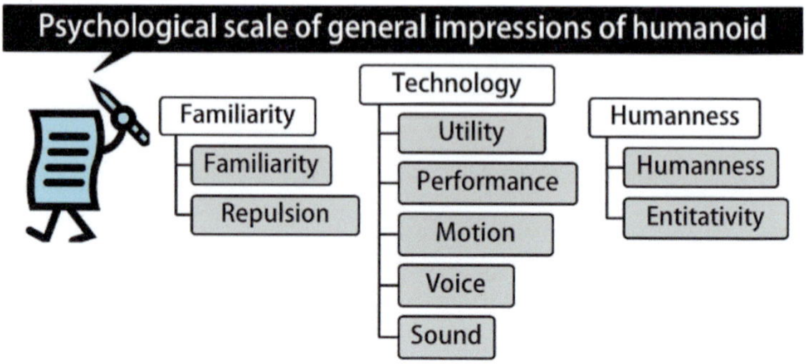

Fig. 8.13 Revealed basic perspectives for the evaluation of humanoid robots

of the humanoid robots on the developed scale to obtain quantitative data (the participants were different from those in Study 1). We kept the distribution of populations of participants the same as in Study 1. We used twice as many participants as we did in Study 1, and the total number was 450 for one presentation experiment and 2700 for all the four experiments. We divided 11 humanoid robots into six groups. The procedure is identical with that of Study 1.

8.4.3.1 Results and Discussion

We conducted an exploratory factor analysis with promax rotation and found nine factors (Fig. 8.13). The first factor called "humanness" includes the humanlike (not robot-like) quality of appearance or facial expressions. The second factor named "motion" captures the clumsiness or smoothness of robot's motion. The third factor named "familiarity" reflects how amiable a robot is or how intimate people feel toward it. The fourth factor named "utility" captures the usefulness and clear goals regarding the functions of robot. The fifth factor named "performance" concerns the high cognitive or technological functions as instruments. The sixth factor named "repulsion" mentions hatred or anxiety toward the existence of robots. The seventh factor named "agency" reflects the extent to which robots has its own mind or intention. The eighth factor named "voice" is the ease with which robot's language can be understood, and the last ninth factor named "sound" is the degree of disturbance in the sound robots make when they move.

The result of the factor analysis was similar to the result of Study 1. That is, Study 1 and Study 2 revealed that nine factors reflect the basic general standards for evaluating humanoid robots, by both qualitative and quantitative approaches. The reliability of each subscale, that is, the consistency of the set of items, is almost sufficiently high (0.73–0.91) except for the factor of voice (0.58).

By using this scale, we can objectively find differences in psychological impressions about different robots. Studies of social backgrounds or cultures are helpful for developing robots that fit situations or persons, and our research is an example. Moreover, we can find not only what parameters are important for psychological impressions but also how parameters are related to impressions. In future work, multiple approaches including social psychological research and engineering development will be needed to investigate whether humanoid robots can indeed coexist with humans.

8.4.4 *Effects of Human Factors of Sex and Age on General Psychological Impressions of Humanoid Robots*

While it is important to develop technologies to improve robot systems by engineering approach, human factors also have implications for the realization of humanoids that can companionably coexist with humans. To investigate the effects of individual differences on the evaluation of humanoid robots, Nomura et al. (2006) developed a psychological scale that focused on the individual tendencies that prevented people from interacting with robots. It is important to discover the effects of human's basic attributes, such as age and sex, on their evaluation of robots, although other characteristics might affect individual tendencies with regard to the relationship with robots.

In this section, our objective is to discover the effects of the age and sex of individuals on how they tend to evaluate humanoid robots on the basis of general impressions about them. The previous section has revealed that there are nine dimensions to evaluate humanoid robots generally, and familiarity, utility, and humanness seem to be three relatively basic dimensions. Hence, we focus on psychological impressions along these three dimensions. In addition, in line with the previous studies, the effect of robot type is also taken into account. There are many criteria for dividing humanoid robots into types. The most apparent criteria from the ordinary viewpoint are how they look and how they move. Therefore, we divide humanoid robots into three types: wheeled-walking robots, biped-walking robots, and android robots. We will report our investigation into the effects of sex, age, and robot types on general impressions of humanoids.

The data was identical with that used in the previous section. We divided the samples into four groups according to age, from the oldest to the youngest; each group consisted of 25 % of the participants and was labeled as follows: we named the 14–28-year-old group "adolescent," the 29–44-year-old group "young age," the 45–59-year-old group "middle age," and the 60–79-year-old group "old age." We applied a three-way ANOVA test to the scores for familiarity, utility, and humanness, in order to investigate the effects of sex (male/female), age (adolescent/young age/middle age/old age), and robot types (wheeled walking/biped walking/android).

Older females judged humanoid robots to be more familiar than females of younger age and males did (Fig. 8.14). Adolescents judged humanoid robots and especially android robots to be less useful. All the groups gave high evaluation of

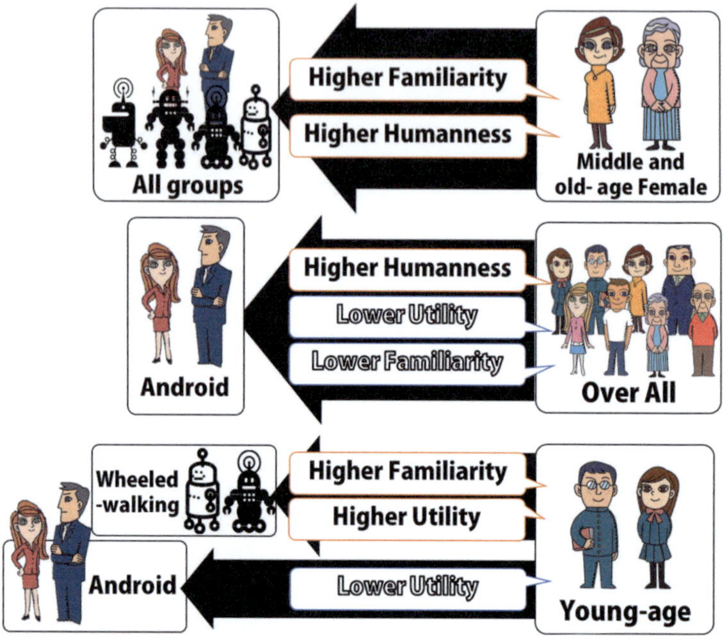

Fig. 8.14 Relationships between human characteristics (sex and age) and general impressions of humanoid robots

humanness to android robots. Our results suggest that it is difficult to make younger people accept humanoid robots or to motivate them to use humanoid robots. This is a potential problem for the marketing and commercial use of humanoid robots. As compared with older people, younger people may be more familiar with the technical aspects of robotic products or more practically oriented. This may be a reason for them to tend to give lower evaluation to humanoid robots generally. On the other hand, older females may care about how humanoid robots look like or how adorable they are, as opposed to their technical aspects. Older females could be more eager users of humanoid robots in the future, by comparison with younger people.

As for the results concerning humanness, android robots had highest evaluation; all the groups evaluated android robots low in familiarity, and hence it seems that android robots were not accepted generally. But older females regarded android robots to be more humanlike than the other robot types. The results concerning humanness and familiarity, then, suggest that older females are more open to humanoid robots including android robots. In sum, females tend to have strong drive to be intimate with others than other age groups, and older people may not be very practically oriented or concerned with the technical aspects of robots.

Acknowledgments We would like to express our appreciation to researchers who provided us with movies of robots and checked them: Satoshi Koizumi (Robovie), Yutaka Nakamura, Ryota Hiura (Wakamaru), Kazuhito Yokoi (HRP-2, HRP-4C), Minoru Asada, Hiroshi Ishiguro, Takashi Minato, and Yuichiro Yoshikawa (Neony, CB2, Repliee Q2, Geminoid, Geminoid F).

Exercises

1. How can service robots contribute to aged society? State two services to be provided by robots and then discuss engineering issues concerning each service.
2. When a humanoid robot performs a pushing task, what is an effective measure to increase its pushing force?
3. How can "familiarity" of service robots be defined? Clarify what characteristics familiarity has and then discuss the specific method for evaluating it.

References

Bartneck, C., Kulic, D., Croft, E., Zoghbi, S.: Measurement instruments for the anthropomorphism, animacy, likeability, perceived intelligence, and perceived safety of robots. Int. J. Soc. Robot. **1**(1), 71–81 (2009)

Harada, K., Hirukawa, H., Kanehiro, F., Fujiwara, K., Kaneko, K., Kajita, S.: Dynamical balance of a humanoid robot grasping an environment. In: Proceedings of the IEEE International Conference on Robotics and Automation, pp. 1167–1173 (2004)

Inoue, K., Ujiie, Y., Takubo, T., Arai, T.: Psychological evaluation for rough shape of humanoid robots using virtual reality. In: Proceedings of the 13th International Conference on Advanced Robotics, pp. 1005–1010 (2007)

Kajita, S.. Kanehiro, F., Kaneko, K., Fujiwara, K., Harada, K., Yokoi, K., Hirukawa, H.: Resolved momentum control: Humanoid motion planning based on the linear and angular momentum. In: Proceedings of the IEEE/RSJ International Conference on Intelligent Robots and Systems, pp. 1644–1650 (2003)

Kamide, H., Takubo, T., Ohara, K., Mae, Y., Arai, T.: Impressions of humanoids: the development of a measure for evaluating a humanoid. Int. J. Soc. Robot. **6**(1), 33–44 (2013)

Negi, S., Arai, T., Inoue, K., Ujiie, Y., Takubo, T.: Psychological assessment of humanoid robot – appearance and performance using virtual reality. In: Proceedings of the 17th IEEE International Symposium on Robot and Human Interactive, pp. 719–724 (2008)

Nomura, T., Suzuki, T., Kanda, T., Kato, K.: Measurement of Anxiety toward robots. In: Proceedings of the 15th IEEE International Symposium on Robot and Human Interactive Communication, pp. 372–377 (2006)

Population statistics, National Institute of Population and Social Security Research. http://www.ipss.go.jp/

Takenaka, T., Hasegawa, T., Matsumoto, T.: Posture control device for leg type mobile robot. Japan Patent Office 10–230485, 1998

Takubo, T., Inoue, K., Arai, T.: Pushing an object considering the hand reflect forces by humanoid robot in dynamic walking. In: Proceedings of the IEEE International Conference on Robotics and Automations, pp. 20051718–1723 (2005)

Vukovratovic, M., Borovac, B., Surdilovic, D.: Zero-moment point – proper interpretation and new applications. In: Proceedings of IEEE-RAS International Conference on Humanoid Robots, pp. 237–244 (2001)

The White Paper on Police: National Police Agency. http://www.npa.go.jp/english/index.htm (2010)

World Robotics: IFR. http://www.worldrobotics.org/ (2007)

Chapter 9
Android Science

Hiroshi Ishiguro

Abstract This chapter discusses android science as an interdisciplinary framework bridging robotics and cognitive science. Android science is expected to be a fundamental research area in which the principles of human-human communications and human-robot communications are studied. In the framework of android science, androids enable us to directly exchange bodies of knowledge gained by the development of androids in engineering and the understanding of humans in cognitive science. As an example of practice in android science, this chapter introduces geminoids, very humanlike robots modeled on real persons, and explains how body ownership transfer occurs for the operator of a geminoid. The phenomenon of body ownership transfer is studied with a geminoid and a brain-machine interface system.

Keywords Human-robot interaction • The behavior-based system • Distributed cognition • Constructive approach • Android science • Total Turing test • Total intelligence • Android • Uncanny valley • Geminoid • Body ownership transfer • Rubber hand illusion • Brain-machine interface • Brain-machine interface (BMI) system

9.1 Introduction

9.1.1 A New Issue in Robotics

In the past, the major topics in robotics were *navigation* and *manipulation*.[1] Robots moved in static environments and manipulated rigid objects. Recently, a new research topic—*human-robot interaction*—has become the focus. The new research goal is to develop robots that can interact with and assist people in everyday

[1]The materials of Sects. 9.1 and 9.2 are adopted and modified from Ishiguro (2007; doi: 10.1177/0278364907074474), thanks to Sage Publications.

H. Ishiguro (✉)
Graduate School of Engineering Science, Osaka University, Toyonaka, Osaka, Japan
e-mail: ishiguro@sys.es.osaka-u.ac.jp

M. Kasaki et al. (eds.), *Cognitive Neuroscience Robotics A*,
DOI 10.1007/978-4-431-54595-8_9

environments. This section discusses the interaction between humans and robots—especially very humanlike robots called "androids" and "geminoids."

9.1.2 *Intelligence as Subjective Phenomena*

How can we define intelligence? This fundamental question motivates researchers in artificial intelligence and robotics. Early works in artificial intelligence assumed that the functions of memory and prediction would realize the intelligence of artificial systems. After the big wave of artificial intelligence in the 1980s and 1990s, researchers faced the difficulty of this approach. In order to find a break though, they focused on the importance of embodiment and started to use robots. The behavior-based system proposed by Brooks (1991) was a trigger of this new wave of artificial intelligence. That is, the focus in artificial intelligence and robotics has shifted from internal mechanisms to interactions with the environment.

On the other hand, there are two approaches to the understanding of human intelligent behaviors in cognitive science. One is to focus on the internal mechanism and the other is to focus on the interactions among people. This latter approach has been studied in the framework of distributed cognition (Hollan et al. 2000). The ideas of distributed cognition and behavior-based system share important aspects. The most important common aspect is to understand intelligence through human-human or human-robot interactions.

This chapter pursues the ideas of behavior-based system and distributed cognition. If we consider that intelligence is a subjective phenomenon, it is important to implement rich interactive behaviors in a robot in order to make it intelligent. We believe that the development of rich interactions among robots will provide clues for the principles of and design methodology for communication systems.

9.1.3 *A Neglected Issue*

When a robot interacts with people, it makes certain impressions on people. In the study of interactive robots, subjective impressions are a common performance measure (Kanda et al. 2001). In addition, interactive robots can elicit largely subconscious reactions, such as eye movements and posture synchronizations (Itakura 2004).

One neglected issue is the appearance of the robot itself. The interactive robots developed thus far are humanoids that encourage interlocutors to anthropomorphize them. Evidently, the appearance of a robot influences people's impressions, and it is a very important factor in its evaluation. Many technical reports have compared robots that have different behaviors; however, the appearance of robots has thus far received scant attention. Although there are many empirical discussions on very simple static figures, such as dolls, the design of a robot's appearance—in particular, the issue of how to make it look lifelike, sociable, or humanlike—has always been a task for industrial designers. This is a serious problem for developing and evaluating interactive robots as well. Appearance and behavior are closely linked to each other, and people's evaluation of a robot changes according to its appearance.

9.1.4 Constructive Approach in Robotics

There are explicit evaluation criteria for robot navigation, such as speed, precision, etc. Our purpose is not just to evaluate but also to develop interactive robots. For this purpose, first, we need to have enough knowledge of humans and develop explicit evaluation criteria for human behaviors.

However, this knowledge is not sufficient to provide a top-down design; instead, a potential approach is bottom-up. By utilizing available sensors and actuators, we can design behaviors of a robot and then decide on the execution rules for selecting the relevant behaviors. In this development of robot design, we also evaluate the robot's performance and modify its behaviors and execution rules. We call this bottom-up approach the "constructive approach" (Ishiguro et al. 2002).

In the constructive approach, cognitive scientists, psychologists, and robotics researchers work together for evaluating and analyzing robots; and then, they improve the robots by using the knowledge obtained through their discussions.

9.1.5 Appearance and Behavior

In the evaluation of robots, the performance measures are subjective impressions (Kanda et al. 2001) and their unconscious reactions, had by human subjects who interact with robots. The unconscious reactions include synchronized human behaviors and eye movements in the interactions (Itakura 2004).

Obviously, both the appearance and behavior of the robots are important factors in this evaluation. Appearance and behavior tightly couple each other and influence each other in the evaluation. We have developed several humanoids that are capable of communicating with people (Ishiguro et al. 2002, 2001; Kanda et al. 2004), as shown in Fig. 9.1, and empirically know the effects of different appearances. They are as significant as behaviors in communication.

9.1.6 Android Science

To tackle the problem of appearance and behavior, a framework including two approaches is necessary: one approach is from robotics and the other is from cognitive science. The approach from robotics tries to build very humanlike robots on the basis of knowledge from cognitive science. The approach from cognitive

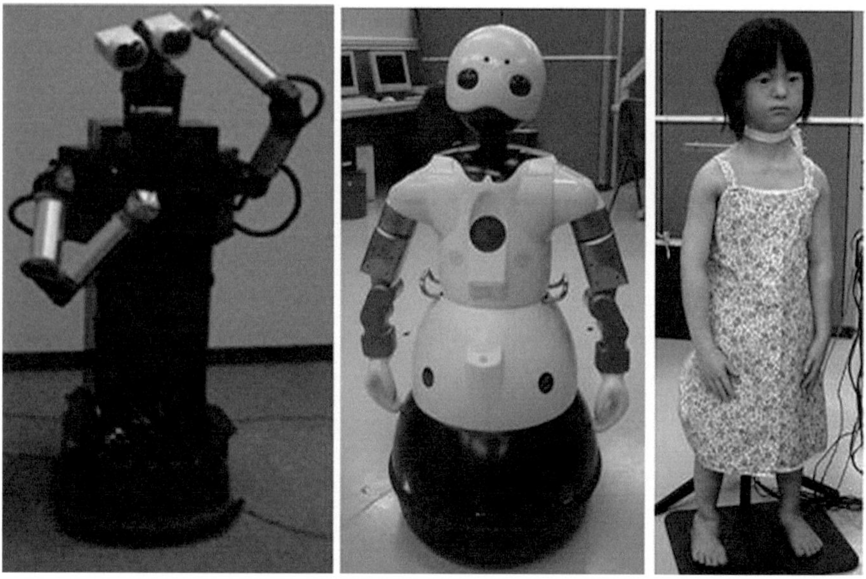

Fig. 9.1 From humanoids to androids. The *left picture* is Robovie II developed by ATR Intelligent Robotics and Communications Laboratories. The *middle* one is Wakamaru developed on the basis of Robovie by Mitsubishi Heavy Industry Co., Ltd. The *right* one is a child android

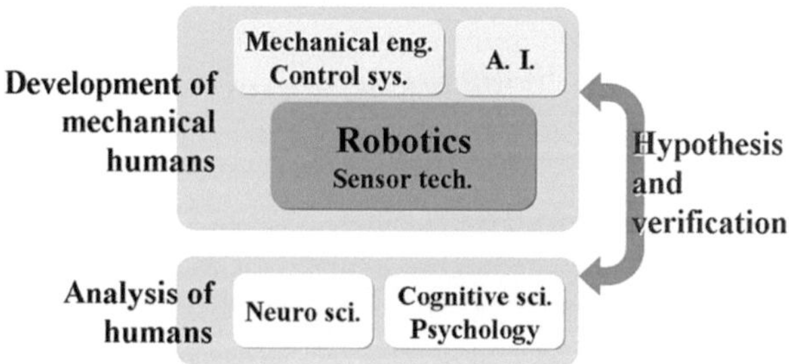

Fig. 9.2 Framework of android science

science uses robots to verify hypotheses for understanding humans. We call this cross-interdisciplinary framework *android science*.

In the past, robotics used knowledge of cognitive science and cognitive science utilized robots. However, the past contribution of robotics to cognitive science was not enough, since robot-like robots were not sufficient tools for cognitive science. We expect that androids that have identical appearances as humans can solve this problem. By using a very humanlike android to study human-robot interaction, we can clearly understand whether certain effects come from its humanlike appearance or from both its humanlike appearance and behavior.

In the framework of android science, as it is shown in Fig. 9.2, androids enable us to directly exchange bodies of knowledge gained by the development of androids in engineering and the understanding of humans in cognitive science. This chapter discusses android science from the perspectives of both robotics and cognitive science.

9.1.7 Conscious and Subconscious Recognition

The goals of android science are to create a humanlike robot and to find the essential factors to replicate *humanlikeness*. How can we define humanlikeness? Furthermore, how do we perceive humanlikeness?

The meaning of humanlikeness remains rather superficial for the time being. Obviously, current technologies cannot produce a perfect android, and the capacity for humanlike interaction is very limited. However, it is possible to develop an android that gives us an impression of humanlikeness during a short-term interaction.

It is well known that human forms elicit various conscious and subconscious responses. A human participant responds to an android in the same way as he/she responds to a human. This constitutes at least a kind of subconscious recognition of the humanlikeness of the android, even if the participant consciously recognizes that the android is not a human. This fact provides us with a conceptual scheme of the cognitive system shown in Fig. 9.3. When we observe objects, various modules are activated in our brains. Each of these modules reads sensory inputs and triggers reactions. An important point here is that the reactions come from modules associated with focal conscious and nonconscious processing. For example, even if we recognize a robot as an android with focal conscious modules, subconscious recognition modules may generate reactions to it as if it were a human. This point is fundamental both for engineering and scientific studies. It offers an evaluation criterion in the development of an android and provides cues for understanding the human brain mechanism for recognition.

9.2 Development of Androids

9.2.1 Humanlike Appearance

The main difference between mechanical-looking robots and androids is their appearances. The appearance of an android is realized by making a copy of an existing person. Besides ours, there are a few related works on android making. One of them is the android heads developed by David Hanson (www.han-sonrobotics.com). He has also developed Albert HUBO that places the head of Albert Einstein on the body of the robot HUBO, a robot developed by Jun-Ho Oh and his laboratory at KAIST (Korea Advanced Institute of Science and Technology). In addition,

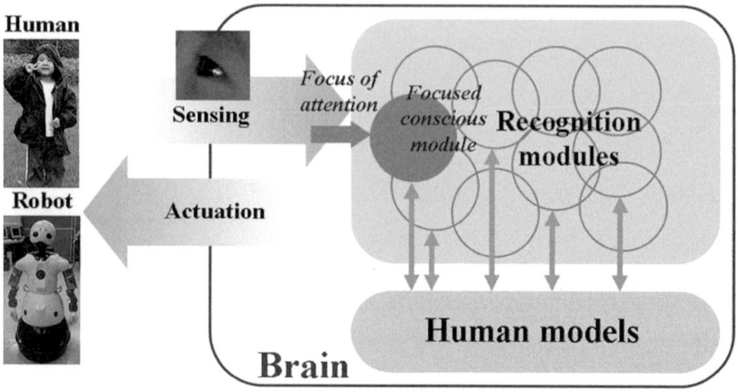

Fig. 9.3 Conscious and subconscious recognition in the human brain

recently, KITECH (Korea Institute of Industrial Technology) has developed the whole body android named "EveR-1."

The author developed an android head and its body before their works. Although there are many androids in historical museums, they do not have a sufficient number of actuators for generating humanlike behaviors and sensors for humanlike perception. In this sense, our android is the first humanlike one in the world, despite its limitations.

The process of making a human copy is the following. First, molds of human body parts are made from a real human using shape memory form (the same form as often used by dentists). Then plaster models of the human parts are made by using the molds. A full-body model is built by connecting the plaster models. Again, a mold of the full-body model is made from the plaster model, and a clay model is made by using the mold. At this point, professionals of formative art modify the clay model without losing its detailed texture. The human body is very soft, and the full-body model cannot maintain the exact shape of human body in the molding process. Therefore, this modification is necessary; it must be done very carefully by referring to photos of the model person. After the modification, a plaster full-body mold is made from the modified clay model, and then a silicone full-body model is made from the plaster mold. This silicone model is used as a master model.

A silicone skin for the full body is made from the master model. The thickness of the silicone skin is 5 mm in our trial manufacture. The mechanical parts, motors, and sensors are covered with polyurethane and the silicone skin. Figure 9.4 shows the silicone skin, inside mechanisms, and head part of a child android, and Fig. 9.5 shows the same android as it is completed. The skin is completed when it is painted in such a way that it closely resembles a human skin, as shown in the figure. It is difficult to distinguish the completed android from the model person merely by looking at their photographs.

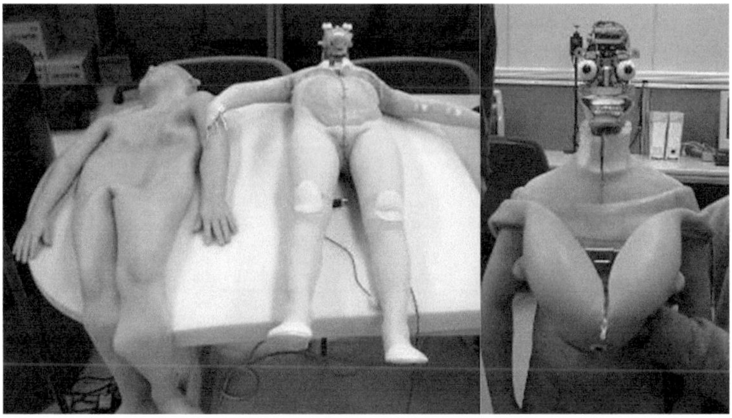

Fig. 9.4 The silicone skin and internal mechanisms

Fig. 9.5 The child android

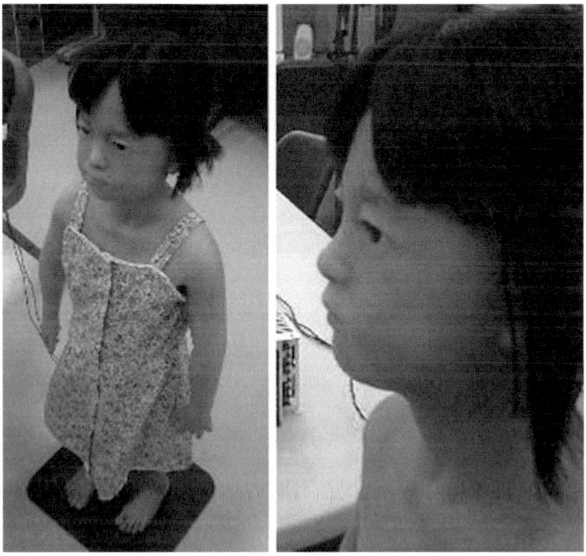

The technology for recreating a human as an android has been accomplished to some extent. However, it is not perfect yet. The major difficulties consist in making the following:

- Wet eyes
- More flexible and robust skin material

The most sensitive part of a human is the eyes. When we face a human, we first look at the eyes. Although the android has a mechanism for blinking and its eyeballs are perfect copies, we are aware of the differences between it and a real

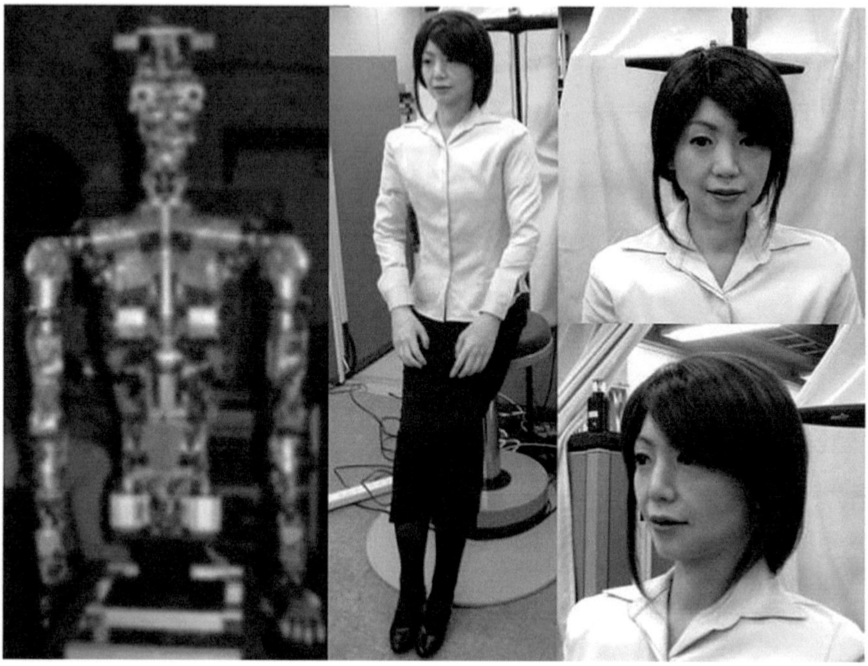

Fig. 9.6 Adult android developed in cooperation with Kokoro Co. Ltd

human. The wet surface of the eyes and the outer corners are difficult to recreate by using silicone. More improvements are required in the manufacturing process.

The silicone currently used in this trial manufacturing is sufficient for recreating the texture of the skin. However, it loses flexibility after 1 or 2 years, and its elasticity is insufficient for large joint movements.

9.2.2 Inside Mechanisms of the Android

Very humanlike movement is another important factor for developing androids. In order to realize humanlike movements, the author developed an adult android; the child android was too small to install a sufficient number of actuators in it. Figure 9.6 shows the developed adult android. The android has 42 air actuators for the upper torso except fingers.

Figure 9.7 shows the arrangement of the major actuators. The android consists of rotary pneumatic actuators and linear pneumatic actuators:

Rotary pneumatic actuator

Hi-Rotor, Kuroda Pneumatics Ltd.
(http://kuroda-precision.co.jp/KPL/english/index.html)

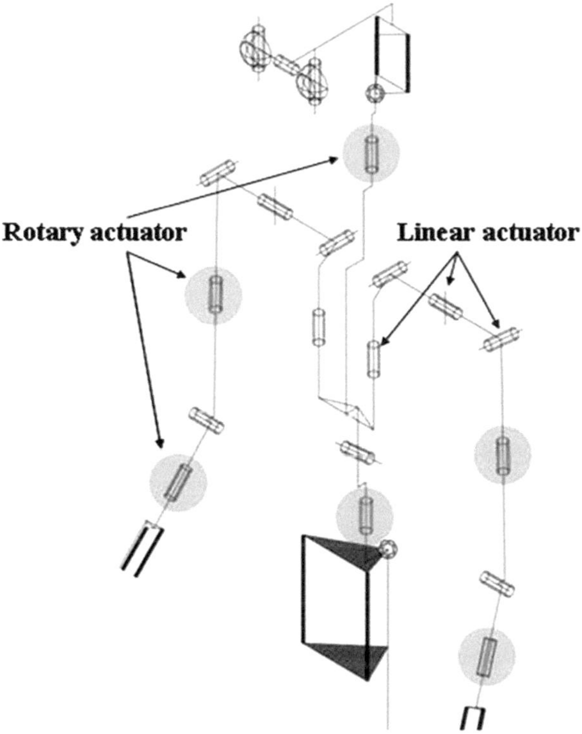

Fig. 9.7 The arrangement of actuators

Linear pneumatic actuator

Air Cylinder, SMC Corporation
(http://www.smcusa.com/)

The specifications of the actuators are as follows. The output power can be decided by the diameter of the air cylinder and the air pressure provided by the air compressor. The smallest linear pneumatic actuator used for the android is

SMC CJ2B16-10-DCH185FH	
Diameter	16 mm
Stroke	10 mm
Maximum operating pressure	0.7 M Pa
Maximum output	140 N

The rotary pneumatic actuator used for the arm is

KURODA PRNA10S-120-45-204-0850	
Maximum operating pressure	0.7 M Pa
Maximum output	140 N

These pneumatic actuators require an air compressor. The system uses the air compressor specified below. One of the key issues in the development is to make the air compressor small or replace pneumatic actuators with electric actuators. The air compressor significantly limits the mobility of the android. It is

LSP-371CD, ANEST IWATA Corporation (http://www.anest-iwata.co.jp/english/index.html)	
Output	0.8–1.0 M Pa
Maximum air weight flow rate	345 l/min
Noise	50 dB A
Normal power	3.7 kW
Dimensions	545 × 663 × 1227 mm

The author and his colleagues adjusted the positions of the actuators by capturing the movements of a real human with a precise 3D motion capturing system. It was difficult to develop an automatic process for this adjustment. Therefore, first, we decided as to what human movements should be represented by the android. It is impossible for the android to perfectly imitate human movements since the actuators are different from human muscles. We focused on small but natural body movements. At the same time, we decided about the number of actuators and their coarse arrangement on the basis of our experience. Then, we adjusted the arrangements by referring to the captured motion data. This was try-and-error adjustments by hand. As a result, we obtained the arrangement that can represent unconscious movements (such as chest movements by breathing) and conscious large movements (such as movements of the head and arms).

Furthermore, the android has functions to generate facial expressions that are important for interaction with humans. Figure 9.8 shows several examples of the android's facial expressions. The android uses 13 of the 42 actuators to make a facial expression. There are a few works on this. For example, Hashimoto et al. have developed an android head for studying human facial expressions (Hashimoto et al. 2006). The android head has 18 actuators arranged for roughly imitating the human muscle structure on the face, and these actuators pull partial areas of the face. The actuators also work to push the skin. Pushing the silicone skin from the back can generate delicate facial movements. The exact arrangements of these actuators are the most important expertise of our collaborating company for making high-quality androids.

There are reasons why we use pneumatic actuators for androids. First, DC servomotors require several reduction gears and hence make un-humanlike noise.

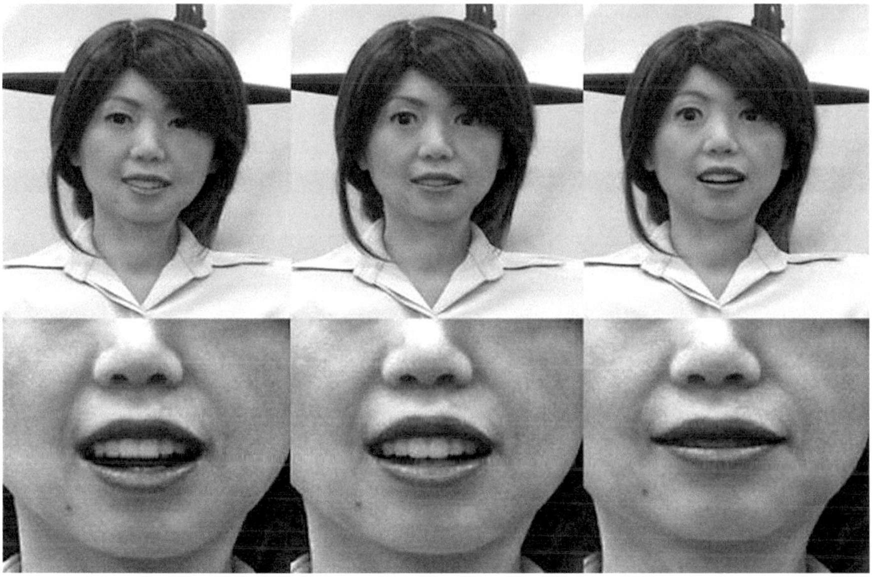

Fig. 9.8 Facial expressions of the android

However, pneumatic actuators are silent enough. Second, the reaction of the android against external force is natural if it is given an air dumper. Although DC servomotors with reduction gears need sophisticated compliance control, pneumatic actuators do not need it. This is important for realizing safe interaction with humans.

However, the drawback of air actuators is that they require a large and powerful air compressor. Because of its big air compressor, the current android model cannot walk. To achieve wide applications, it is necessary to develop new electric actuators that have characteristics similar to air actuators.

9.2.3 *Humanlike Movements*

The next issue is how to control 42 pneumatic actuators for realizing humanlike natural movements. Here, there is a difference between our android and other robots. The pneumatic actuator works as a damper. Therefore, if we move one actuator, the surrounding actuators also move. That is, we have to control almost all actuators simultaneously.

The simplest approach for making humanlike movements is to directly send angular information to each joint by using a simple graphical user interface termed a "motion editor." By using the motion editor, we register postures of the android, and then it interpolates the postures and generates continuous motion commands to the actuators. If we patiently register the posture, this method represents any kind of movement.

Of course, easier functions are preferred, and therefore we, the author and his colleagues, are improving the motion editor. One idea is to approximate the joint movements with smooth functions on the basis of sinusoidal signals. This is the same idea as Perlin noise used in computer graphics (Perlin 1995). This idea works very well for particular movements, for example, arm movements, but it is rather limited. When body movements are represented as combinations of sinusoidal curves, they are often very complex, and any simple method fails to handle them.

A more advanced method for generating humanlike natural movements is to imitate directly the movements of the operator. However, the android made from the silicone skin faces specific difficulties. In addition to the complexity arising from the many DOFs, a difficulty is that the skin movement does not simply correspond to the joint movement. For example, the android has more than five actuators around the shoulder for generating humanlike shoulder movements, and the skin moves and stretches as the actuators move. Therefore, even if we know the mechanical structure of the android, we cannot precisely control skin movements.

The idea for overcoming this difficulty is to train a neural network. Figure 9.9 shows the experimental setup. In the figure, the android and its original person have makers of the 3D motion capture system on the identical positions. The feedback for training the neural network is the sum of differences in the marker positions. Since we do not know the exact mapping between actuator movements and skin movements, we need to randomly search for the actuator movements that minimize the error. This search is very time consuming, and so we have employed a simulated annealing technique for saving the searching time.

However, even this technique takes long time. It is necessary to improve the training method. In addition, the current implementation is a position-based control. It works for very small and slow movements but not for large and quick movements. We may be able to control the speed and torque. But then, the difficulty will be more complicated.

It is left to future work in android development to fully overcome the difficulty. Nevertheless, the past attempts suggest an interesting topic of study. The imitation by the android means representing the complicated human shape and motions in the parameter space of the actuators. We may find important properties of human body movements by analyzing the parameter space. More concretely, we expect to have a hierarchical representation of human body movements that consist of two or more layers, such as small unconscious movements and large conscious movements. This hierarchical representation provides flexibility in android behavior control.

9.2.4 Humanlike Perception

The android requires humanlike perceptual abilities, in addition to humanlike appearance and movements. This requirement has been tackled by computer vision and pattern recognition techniques in controlled environments; however, it becomes

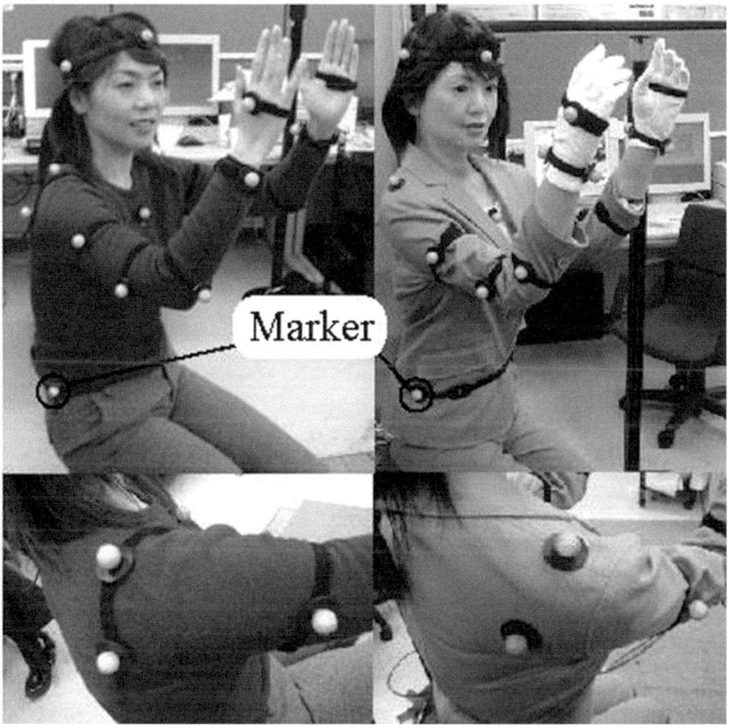

Fig. 9.9 Positions of motion capture makers on the original person (*right*) and the android (*left*)

very difficult to meet the requirement in real and messy situations; vision becomes unstable, and audition and speech recognition is hampered by background noise.

The distributed sensor systems of ubiquitous computing may solve this problem. The idea is to recognize the environment and human activities by using many distributed cameras, microphones, infrared motion sensors, floor sensors, and ID tag readers in the environment. In our previous work, we have developed distributed vision systems (Ishiguro et al. 1997) and distributed audition systems (Ikeda et al. 2004). This work must be integrated and extended to solve the problem. Figure 9.10 shows the sensor network installed in the laboratory. The omnidirectional cameras observe humans from multiple viewpoints and robustly recognize their behaviors (Ishiguro et al. 2001). The microphones capture human voice by forming virtual sound beams. The floor sensors that cover the entire space of the laboratory reliably detect footsteps.

Skin sensors are often installed on robots. Soft and sensitive skin sensors are particularly important for interactive robots. However, there is no skin sensor to sufficiently mimic human sensitive and soft skin. We have developed novel skin

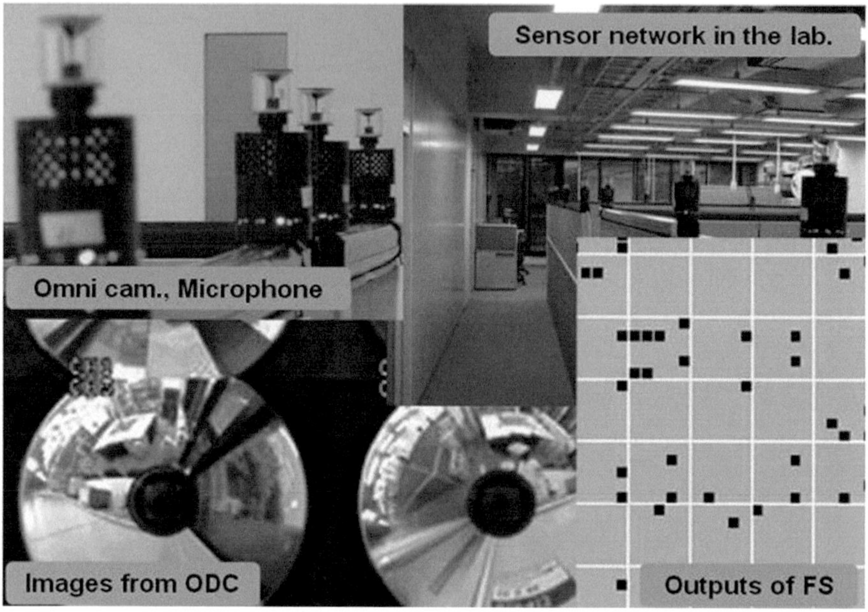

Fig. 9.10 Distributed sensor system

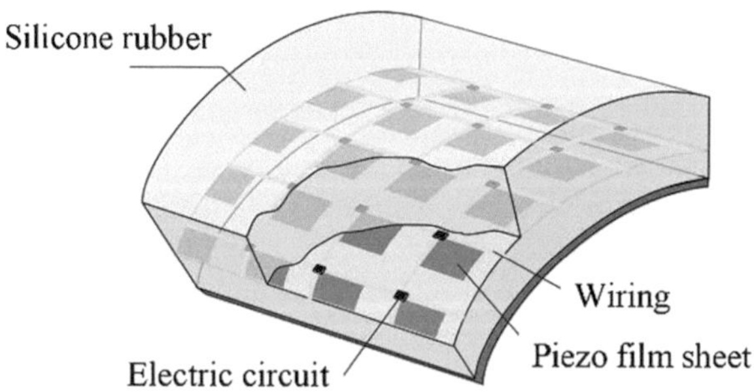

Fig. 9.11 Skin sensor

sensors for our androids. The sensors are made by combining silicone skin and piezo films, as shown in Fig. 9.11. These sensors detect pressure from the bending of piezo films. By increasing their sensitivity, they can detect the presence of an object in the immediate vicinity, because of static electricity. These technologies for humanlike appearance, behavior, and perception enable us to realize the humanlike androids.

9.3 Cognitive Studies with Androids

9.3.1 Total Turing Test

As discussed above, the framework of android science includes two approaches: the engineering approach and the scientific approach. The most typical experiment where the two approaches meet is the total Turing test. The original Turing test is devised to evaluate the intelligence of computers, under the assumption that mental capacities can be separated from embodiment (Turing 1950). This assumption has invoked many questions about the nature of intelligence. We regard intelligence as subjective phenomena among humans or between a human and a robot. Obviously, the original Turing test does not cover the concept of total intelligence (Harnad 1990). By contrast, androids enable us to evaluate total intelligence.

As the original Turing test does, the total Turing test uses a time competition. We, the author and his colleagues, did a series of preliminary experiments to check how many people do not become aware of the fact that they are facing an android. Figure 9.12 shows the experimental scene. The task for subjects is to find the color of the android's cloths. The screen between the android and the subject opens for 2 s. The subject then identifies the color. The experimenter asks the subject whether he/she has become aware that it is an android. We have prepared two types of androids: one is a static android and the other is an android with micro movements that we call "subconscious movements." We, humans, do not freeze; we are always slightly moving even when we are not doing anything in particular, e.g., when we are just sitting on a chair. The android shown in Fig. 9.13 performs subconscious micro movements.

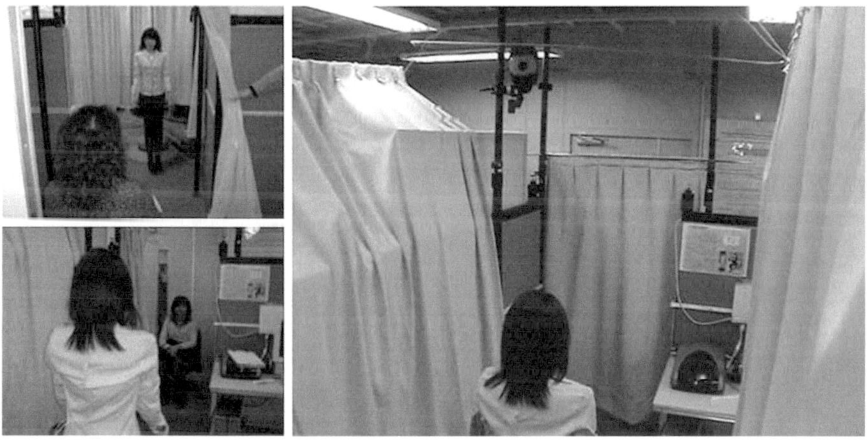

Fig. 9.12 Total Turing test

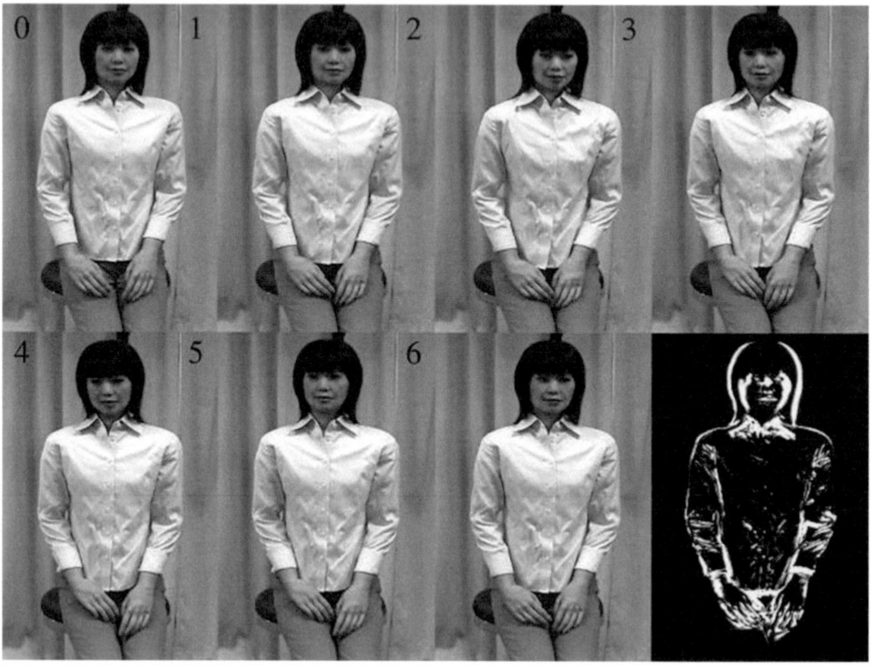

Fig. 9.13 Subconscious micro movements of the android. Seven snapshots and the movements (*bottom right*)

We have prepared three experimental conditions: a static android, a moving android, and a real human. There are two time conditions of 1 and 2 s. Figure 9.14 shows the results; (a) and (b) show the results for the time conditions of 1 and 2 s, respectively. Figure 9.14a reveals that 80.0 % of the subjects become aware of the android when they look at the static android; however, 76.9 % of the subjects do not become aware of the android in the moving condition; instead, they recognize it as a human. We can confirm significant differences among these conditions with $p < 0.5$. This result shows the importance of micro movements for real humanlike appearance.

The two-second experiment does not suggest that the android has passed a total Turing test. Nevertheless, it shows significant possibilities for the android itself and interdisciplinary studies across engineering and cognitive science.

Of course, an important challenge is to extend the time beyond 2 s; and more sophisticated technologies are required for meeting this challenge. For example, it is needed to develop a more natural motion generator that is inspired by the human brain mechanism. Furthermore, humanlike reactions are also required. The android already has sensitive skin and can make simple reactions. However, this is not enough. It is needed to integrate skin sensors into other kinds of sensors for vision and audition.

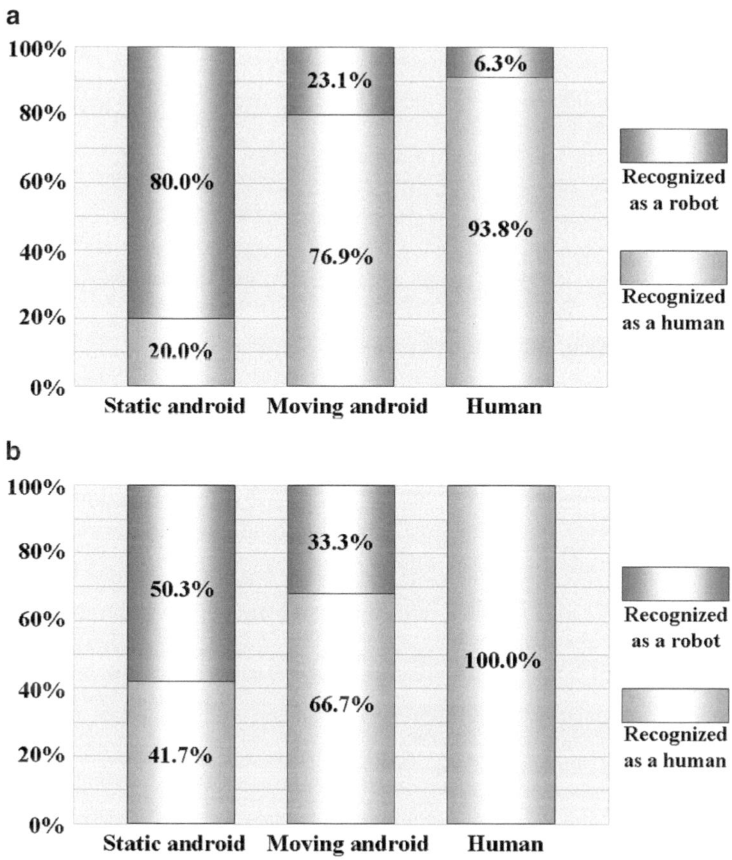

Fig. 9.14 The experimental results (**a**) one-second condition, (**b**) two-second condition

9.3.2 Subconscious Recognition

Another important criterion for evaluating the android is the extent to which it elicits conscious and subconscious social responses. In our experience, people react to the android as if it were human, even when they consciously recognize it as an android. Their reaction to the android is very different from their reactions to robot-like robots. For example, people hesitate to touch the android just as they hesitate to touch human strangers.

To investigate in detail how people's reactions are different, we, the author and his colleagues, have conducted an experiment and observed the eye movements of the participants. Figure 9.15 shows the differences in eye movements between when the participants observe a child and observe a child android. We intentionally use an eerie child android for this experiment. The child android seems eerie because of its jerky movements. As shown in Fig. 9.15, the participants cannot keep gazing at

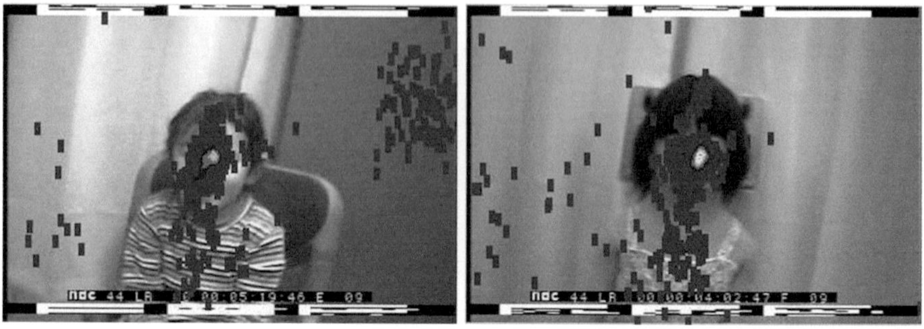

Fig. 9.15 Eye movements with respect to a human child and the android

the face of the human child and often look at the upper right corner. By contrast, the participants keep gazing at the face of the android.

Works in psychology suggest three reasons why the participants break eye contact with a human being under a cognitive load, such as trying to answer a question that demands much thinking:

Arousal reduction theory: Humans shift their gaze direction to eliminate distraction and aid concentration.

Differential cortical activation theory: A downward eye movement is caused by brain activity.

Social signal theory: The eye movement is a means for letting the other person know you are thinking (McCarthy et al. 2001).

The first and second reasons do not match the experimental results. According to the third reason, a human indicates that he/she is social by not gazing continuously at the face of the other. There are differences in the direction of gaze shift among different cultures. Canadians tend to break eye contact by looking up and to the right, while Japanese tend to look down (MacDorman et al. 2005). However, people do not completely stop making eye contact while talking with a person, if they recognize the person as a social partner.

9.3.3 Uncanny Valley

Why do the participants have an eerie feeling when they are looking at the child android? In the experiment, the participants perceive a certain degree of strangeness in the android's movements and appearance. Mori (1970) predicted that as robots appear more human, they seem more familiar until they reach a point at which subtle imperfections create a sensation of strangeness, as shown in Fig. 9.16. He referred to this as the *uncanny valley*.

Fig. 9.16 Uncanny valley

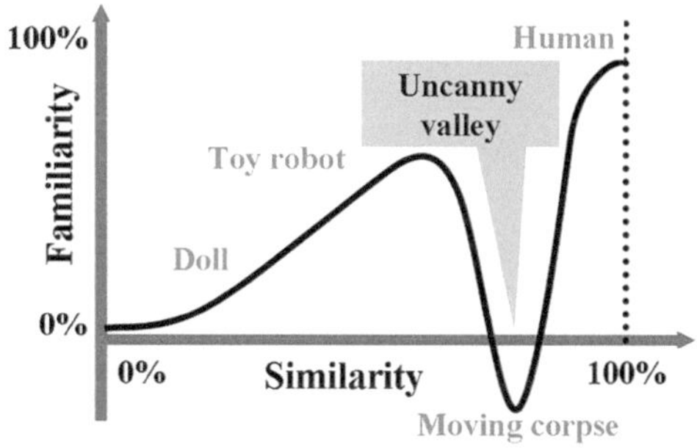

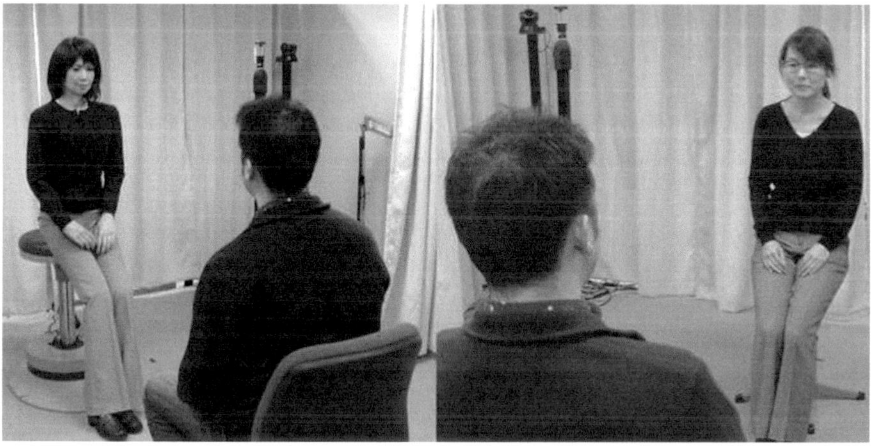

Fig. 9.17 A participant faces the adult android and a human

9.3.4 The Possibility of An Android as a Human

We, the author and his colleagues, have conducted another experiment with an adult android that has a complex mechanism for generating humanlike movements. The experiment involves a 5-min period where a participant is habituated to the android and the android asks questions. During the 5-min habituation period, first, the android introduces itself and makes natural body movements, and then the android asks questions, and the participant answers them. Here, 5 min is sufficiently long to observe the details of the android, and all participants have found that it is not a human. Figure 9.17 shows the experimental scene.

The task for participants is to answer seriously both easy and difficult questions given by the android. In this experiment, almost all participants have shifted their gaze when answering difficult questions. Moreover, we have compared human-human interaction and human-android interaction. Figure 9.18 shows that participants shift their gaze in the same manner when they interact with a human and the android.

Human-Human interaction

2.2	8.3	12.8
10.8	30.2	6.5
4.4	22.7	2.0

Human-android interaction

3.1	3.1	2.0
4.3	18.2	8.2
20.7	19.9	20.5

Fig. 9.18 Comparison between human-human interaction and human-android interaction. The gaze directions are represented with nine areas. The numbers represent the percentages of the gazing directions

It is clear that the participants consciously recognize the human form as an android. Nevertheless, their reactions to it are similar to those to a human in this experiment. This is not enough to conclude that the participants subconsciously recognize the android as a human and treat it as a social partner. However, it opens up a possibility.

9.4 Geminoid: The Teleopeareted Android of an Existing Person

The most difficult task for androids is verbal communication.[2] It is regarded as a bottleneck. We, the author and his colleagues, have developed geminoid, a new category of robot, to overcome this bottleneck issue. We coined "geminoid" from the Latin "geminus," meaning "twin" or "double," and added "oides," indicating "similarity" or "being a twin." As the name suggests, a geminoid is a robot that will work as a duplicate of an existing person. It appears like an existing person and can behave just as he/she does if it is operated by the person via a computer network. Geminoids extend the applicable field of android science. Androids are designed for studying human nature in general. With geminoids, we can study such personal aspects, as human presence or personality traits, and trace their origins and implement them into robots. Figure 9.19 shows the robotic part of HI-1, the first geminoid prototype.

The appearance of a geminoid is based on an existing person and does not depend on the imagination of designers. Its movements can be made or evaluated simply by referring to the original person. The existence of a real person analogous to the robot enables easy comparison studies. Moreover, if a researcher is used as an original person, his/her geminoid counterpart provides meaningful insights for the researcher.

[2]The material of Sect. 9.4 is adopted and modified from Ishiguro and Nishio (2007), with kind permission from Springer Science + Business Media.

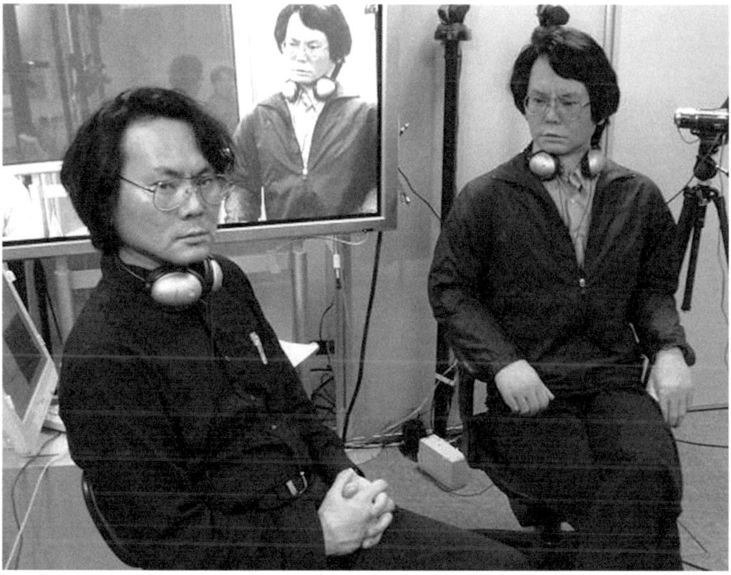

Fig. 9.19 Geminoid HI-1

Since geminoids are equipped with teleoperation functionality, an autonomous program does not fully drive them. The manual control avoids the limitations in current AI technologies and enables us to perform long-term conversational human-robot interaction experiments. This feature also enables us to study human characteristics by separating "body" and "mind." The operator (mind) of a geminoid can be easily changed, while the robot (body) remains the same. Furthermore, the strength of the connection or information transmitted between the body and mind can be easily reconfigured. This is especially important when taking a top-down approach: it adds/deletes certain elements to/from a person to discover the "critical" elements that comprise the person's human characteristics.

The geminoid HI 1 consists of roughly three elements: a robot, a central controlling server (geminoid server), and a teleoperation interface.

The robotic element has essentially identical structure as earlier androids (Ishiguro 2007). However, efforts have been made to build a robot that appears to be a copy of the original person, as opposed to a robot that is merely similar to a living person. Silicone skin is molded by a cast taken from the original person; shape adjustments and skin textures are painted manually on the basis of MRI scans and photographs. Fifty pneumatic actuators drive the robot to generate smooth and quiet movements; this is important when the geminoid interacts with humans. The allocations of actuators are decided in such a way that the resulting robot can effectively show necessary movements for human interaction and simultaneously express the original person's personality traits. Among the 50 actuators, 13 are embedded in the face, 15 in the torso, and the remaining 22 move the arms and

Fig. 9.20 Teleoperation interface

legs. In addition, the softness of the silicone skin and the compliant nature of the pneumatic actuators provide safety for interaction with humans. Since this prototype was aimed for interaction experiments, it lacks the capability to walk around; it always remains seated. Figure 9.19 shows the resulting robot (right) with the original person, Ishiguro, who is the author of this chapter.

Figure 9.20 shows the teleoperation interface of the geminoid. Two monitors show the controlled robot and its surroundings, and microphones and a headphone are used to capture and transmit utterances of the operator. Captured sounds are encoded and transmitted to the geminoid server by IP links between the interface and the geminoid and vice versa. The operator's lip corner positions are measured by an infrared motion capturing system in real time, converted to motion commands, and sent to the geminoid server via the network. This enables the operator to implicitly generate suitable lip movements on the robot while speaking.

Using a simple GUI interface, the operator can also explicitly send commands to control the geminoid's behaviors. Several selected movements, such as nodding, opposing, or staring in a certain direction, can be specified by a single mouse click. We have prepared this relatively simple interface, since the geminoid has 50° of freedom; it is one of the world's most complex robots and its real-time manual manipulation is basically impossible. A simple, intuitive interface is necessary for the operator to be able to concentrate on interaction rather than robot manipulation.

The geminoid server receives robot control commands and sound data from the remote controlling interface and adjusts and merges inputs; in addition, it sends primitive controlling commands to the robot hardware and receives feedback

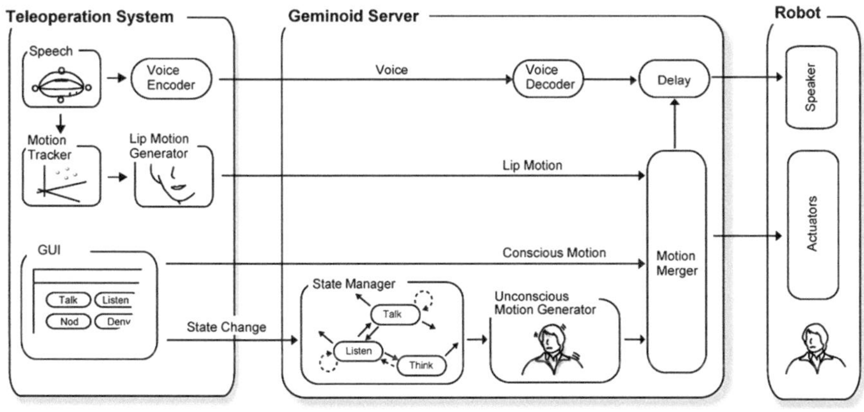

Fig. 9.21 Data flow in the geminoid system

from it. Figure 9.21 shows the data flow in the geminoid system. The geminoid server also registers programmed states of human-robot interaction and generates autonomous or subconscious movements of the robot. Because of the uncanny valley problem, as the robot's features become humanlike, its behaviors should become suitably sophisticated to retain a "natural" look. One thing that every human being performs but most robots fail to do is the slight body movements caused by an autonomous system, such as breathing and blinking. To increase the robot's naturalness, the geminoid server emulates the human autonomous system and automatically generates these micro/subconscious movements, depending on the state of interaction each time. When the robot is "speaking," it shows different subconscious movements than when it is "listening" to others. Such automatic robot motions are generated without the operator's explicit orders but merged and adjusted with conscious operation commands from the teleoperation interface (Fig. 9.21). The geminoid server gives the transmitted sounds a specific delay, taking into account the transmission delay/jitter and the start-up delay of the pneumatic actuators. This adjustment serves for synchronizing lip movements and speech and thus for enhancing the naturalness of movements of the geminoid.

9.4.1 Experiences with the Geminoid

The geminoid HI-1 was completed and press released in July 2006. Since then, numerous operations have been held, including interactions with lab members and experiment subjects. The geminoid was demonstrated to a number of visitors and reporters. During these operations, we, the author and the laboratory members, encountered several interesting phenomena.

When I (Ishiguro, the original person modeled by the geminoid) first saw the HI-1 sitting still, it was like looking in a mirror. However, when it began moving, it looked like somebody else, and I could not recognize it as myself. This means that we, humans, do not objectively recognize our subconscious movements.

While operating the geminoid with the operation interface, I find myself unconsciously adapting my movements to the geminoid's movements. The current geminoid cannot move as freely as I can. I feel that not just the geminoid but also my own body movements are restricted to the geminoid's movements.

When an interlocutor pokes the geminoid, especially around the face, I get a strong feeling that I myself am poked. This is strange since the system currently provides no tactile feedback. Just by watching the monitors and interacting with interlocutors, I get this feeling.

In less than 5 min, both interlocutors and I quickly adapt to conversation via the geminoid. Interlocutors recognize and accept the geminoid as myself. When interlocutors are asked how they felt when interacting through the geminoid, most say that when they saw the geminoid for the first time, they thought that somebody (or Ishiguro, if familiar with me) was waiting there. After taking a closer look, they soon realize that the geminoid is a robot and begin to have some weird and nervous feelings. However, shortly after having a conversation via the geminoid, they find themselves concentrating on the conversation, and the strange feelings have vanished. Most interlocutors are non-researchers who are unfamiliar with robots of any kind.

Does this mean that the geminoid has overcome the "uncanny valley"? Before talking via the geminoid, interlocutors' initial responses to the geminoid seemingly resemble their reactions to other androids; even though they cannot recognize an android as artificial at the beginning of the first meeting with it, they nevertheless soon become nervous in the middle of the meeting. Is intelligence or long-term interaction a crucial factor for overcoming the uncanny valley?

We certainly need some objective means to measure how people feel about geminoids and other types of robots. Minato et al. found that gaze fixation revealed criteria for the naturalness of robots (Minato et al. 2006). Recent studies report on different human responses and reactions to natural or artificial stimuli of the same nature. Perani et al. show that different brain regions are activated while watching human and computer graphic arm movements (Perani et al. 2001). Kilner et al. show that body movement entrainment occurs when watching human motions, but it does not when watching robot motions (Kilner et al. 2003). By examining these findings with geminoids, we may be able to find some concrete measurements of humanlikeness and approach the "appearance versus behavior" issue. The following sections introduce and discuss such research approaches.

9.5 Body Ownership Transfer to Teleoperated Android Robot

9.5.1 Body Ownership Transfer

Through various studies on the geminoid, we have found that the conversation via the geminoid affects not only interlocutors but also its operators.[3] Soon after starting operating the geminoid, they tend to adjust their body movements to the movements of the geminoid. For example, they talk slowly in order to synchronize with the geminoid's lip motion, and they make small movements as the geminoid does.

Some operators even feel as if they themselves have been touched when others touch the teleoperated android (Nishio et al. 2007). For example, when someone pokes the geminoid's cheek, the operators feel as if their own cheek has been poked despite the lack of tactile feedback. This illusion occurs even when other operators than the original persons of geminoids are operating. However, the illusion does not always happen, and it is difficult to induce it deliberately.

There exists a similar illusion named *rubber hand illusion* (RHI) (Botvinick 1998). In RHI, an experimenter repeatedly strokes a participant's hand and a rubber hand at the same time (RHI procedure). The participant can only see the rubber hand and not his/her own hand. After repeating this procedure for a while, the participant begins to have an illusion or feeling as if the rubber hand is his/her own hand. When only the rubber hand is stroked, the participant feels as if her own hand is stroked. This illusion, RHI, is said to occur because of the synchronization between the visual stimulus (watching the rubber hand being stroked) and the tactile stimulus (feeling that the participant's hand is stroked) (Tsakiris 2010). The resulting illusionary effect is quite similar to the effect in our teleoperated android.

However, in the case with the geminoid, this happens without any tactile stimulus; the operator only teleoperates the geminoid and watches it moving and being poked. Our hypothesis is that this illusion—body ownership transfer toward the teleoperated android—occurs due to the synchronization between the operation of the geminoid and the visual feedback of seeing the geminoid's motion. That is, body ownership is transferred by seeing the geminoid moving as the operator moves.

If body ownership transfer can be induced merely by operation without haptic feedback, this can lead to a number of applications, such as realizing a highly immersive teleoperation interface and developing prosthetic hands/bodies that can be used as one's real body parts.

---

[3]The material on body ownership transfer below is adopted and modified from Nishio et al. (2012), with kind permission from Springer Science + Business Media.

We can verify this by comparing cases where participants watch the geminoid in sync with their motion and where they watch the robot out of sync. More concretely, the author and his colleagues have made the following hypotheses and attempted to verify them:

Hypothesis 1 Body ownership transfer toward the geminoid body occurs through synchronized geminoid teleoperation with visual feedback.

RHI requires synchronization of visual and tactile senses. However, the geminoid cannot react immediately, since its actuators put limitations on the geminoid's reaction. In fact, the geminoid usually moves with 200–800 ms delays. We adjust the delaying voice as if the geminoid produces voice in synchronization with mouth movements (Nishio et al. 2007). According to related work, these delays during teleoperation are taken to reduce the extent of body ownership transfer. For example, Shimada et al. show that the RHI effect largely decreases when visual and tactile stimuli differ more than 300 ms in received time (Shimada et al. 2009).

Even if this applies to RHI body ownership transfer, it cannot explain why body ownership is transferred to the geminoid during operation. Therefore, the mechanism of the geminoid body ownership transfer might differ from the mechanism of RHI body ownership transfer. Then, we, the author and his colleagues, have made the next hypothesis:

Hypothesis 2 In geminoid teleoperation, body ownership is transferred when the geminoid moves with delays.

9.5.2 Experiments of Body Ownership Transfer

On the basis of the hypotheses of the previous section and knowledge culled from related studies, we have experimentally verified that body ownership is transferred by the geminoid operation and its visual feedback. When the geminoid is operated, its mouth and head are synchronized with its operator for conversation. However, it is difficult to control the experimental condition if we assign a conversation task to subjects. That is why we set an operation of the geminoid's arm as a task for subjects.

The participants in the experiment included 19 university students: 12 males and 7 females, whose average age was 21.1 years old (standard variation was 1.6 years old). All of them were right handed.

First, the participants operated the geminoid's arm and watched the scene for a fixed time. At this time, the geminoid's arm is synchronized with the operator's arm. This is equivalent to making the participants watch the scene where a rubber hand is stroked in the RHI procedure. Next, we gave the geminoid's arm painful stimuli and measure the self-report and skin conductance response (SCR) of each participant. We predicted that both measurements would show the results that body ownership was transferred.

SCR shows significant values when the autonomic nervous system is aroused, such as when people feel pain (Lang et al. 1993). Armel et al. verified body ownership transfer by measuring SCR (Armel and Ramachandran 2003). Their idea was that if RHI occurred, skin conductance would react when the target object (rubber hand) received a painful stimulus. Participants watched a scene where the rubber hand was bent strongly after the RHI procedure (synchronized/delayed condition). They confirmed that the SCR value in the synchronized condition is higher than in the delayed condition.

The participants in our experiment looked at the geminoid and might believe that it was a real human because it had a very humanlike appearance. That is why the participants were asked to look at it before the experiment and made sure that the geminoid was a nonhuman object.

Then, they learned how to operate the geminoid. At this time, they also checked the camera that was set over the geminoid and learned that they would watch the geminoid through it. However, this camera was not used, as we will discuss later.

After that, they entered the experimental room and worn a marker for the motion capture system on their right arm and electrodes on the left hand. The marker was placed at the position of 19 cm from their elbow to maintain the arm movement radius. They wore electrodes on the ball of the hand. Then, they were told to grasp their hand with the marker, to aim their palm down and aim the other palm up, and not to move their fingers and wrists. They had the same practice as in the main trials. After practicing, they were told about the main trial procedure and questionnaire. Then, they wore a head-mounted display (HMD) and watched two stimulus images: the geminoid's right little finger being bent (the center of Fig. 9.22) and an experimenter injecting the top of the geminoid's right hand (the right of Fig. 9.22). The participants watched these images several times, following Armel and Ramachandran (2003). They re-practiced with the HMD. After this preparation, they challenged the main trials. When all the trials were over, they answered a questionnaire about the experiment.

In the main trials, we repeated the following procedure six times. First, the participants operated the geminoid's right arm by moving their right arm in a

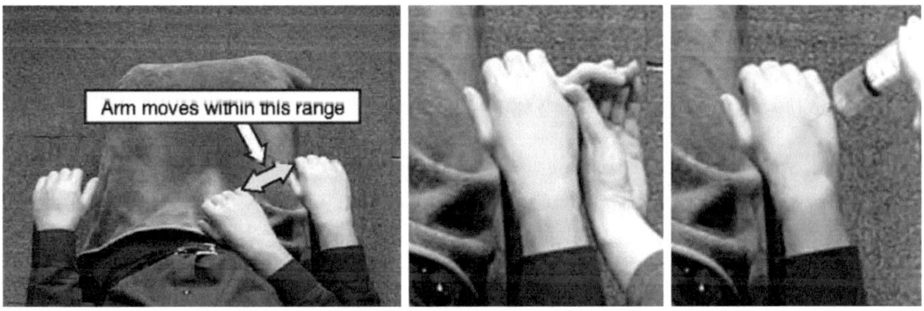

Fig. 9.22 Simulated views shown to participants: (*left*) normal view showing arm movement range; (*center*) "finger bending" stimulus; (*right*) "injection" stimulus

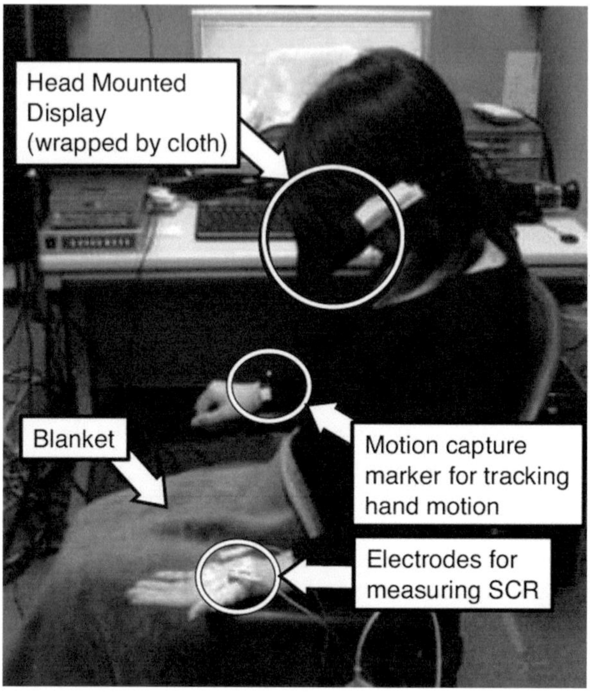

Fig. 9.23 Participant setting

horizontal direction at 3-s intervals. They watched the geminoid's arm movements through the HMD. At this time, they were taught to look down to synchronize their posture with the geminoid's on the HMD because such posture synchronization is important for body ownership transfer. We covered the HMD with a black cloth so that the participants could only see the display. This is because we believe that the extent of body ownership transfer decreases if the participants can see the scene of the experimental room and their body. Moreover, both the participants and the geminoid wore a blanket to prevent their differences in outfit from influencing body ownership transfer. Figure 9.23 shows the participants during the experiment.

After 1-min operation, the experimenters gave the participants a signal for the end and waited until their SCR recovered to normal. Then, the participants were shown one of the two stimulus images again.

In this experiment, the system needs to control the delay on the geminoid. However, it cannot operate the geminoid immediately because of the system limitation. Then, we employed a simulation system. It selected and displayed pictures made from the images of the geminoid hands shot previously. Each picture corresponds to a position of the arm, tracked by a motion capture system. A series of pictures, when displayed continuously, look like a movement of the arm. With this system, we implemented a condition where the geminoid had arbitrary delays, as well as a condition where the geminoid was mostly synchronized with the operator.

We took pictures with a high-speed camera in 300 fps and used 5000 of them for the experiment (the left of Fig. 9.22). We switched the operation images to the stimulus images after the subject's operation.

We measured the participants' self-reports by a questionnaire and SCR. First, we evaluated the extent of body ownership transfer by asking the participants the following questions and examining their self-reports:

(a) Did it feel as if your finger was being bent?
(b) Did it feel as if your hand was being injected?

The participants answered these questions on a seven-point Likert scale (1, not strong; 7, very strong). We made the following three conditions from the hypotheses:

Sync condition: The geminoid movement is synchronized with the operator movement without a delay.
Delay condition: The geminoid movement is synchronized with the operator movement with a certain delay.
Still condition: The geminoid remains still despite the operator's movement.

In the delay condition, the delay was set to 1 s, following Armel and Ramachandran (2003). We conducted six trials to verify the hypotheses: the three conditions by two stimuli, for each participant. The order of the conditions and stimuli was counterbalanced among the participants.

9.5.3 The Experimental Results

Generally, skin conductance reacts with 1–2-s delays for a stimulus. Therefore, the usual method for the verification of the reaction to a stimulus is to measure the maximum value between two points: the stimulus point as the start point and the point at which SCR is calm as the end point. However, we confirmed in this experiment that skin conductance often reacted before the stimulus. The participants answered in the interviews after the main trials that they felt as if they got the stimuli to their hand and felt unpleasant when the experimenter's hand approached the geminoid's hand after the operation. This suggests that body ownership transfer also causes the reaction before the stimulus.

In this experiment, it takes 3 s to give the stimulus (picture of injection) to the geminoid's hand after the experimenter's hand appears. Here, we set the start point at 2 s after the stimulus image starts (1 s before the stimulus is given) and the end point at 5 s after the stimulus is given, as Armel and Ramachandran (2003) did.

The SCR value fulfills normality by a logarithmic transformation. Therefore, we conducted a logarithmic transformation of the SCR value and a parametric test. As a result of one-way ANOVA, no significant difference was confirmed in finger bending ($F(2) = 0.66, p = 0.52$), but a significant difference was confirmed in injection ($F(2) = 3.36, p < 0.05$). As a result of multiple comparisons of Tukey HSD, a significant difference was confirmed only between the sync and still conditions in

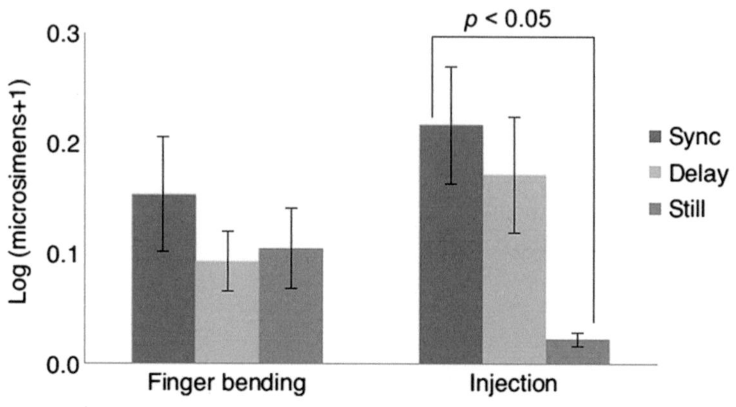

Fig. 9.24 Results of SCR analysis

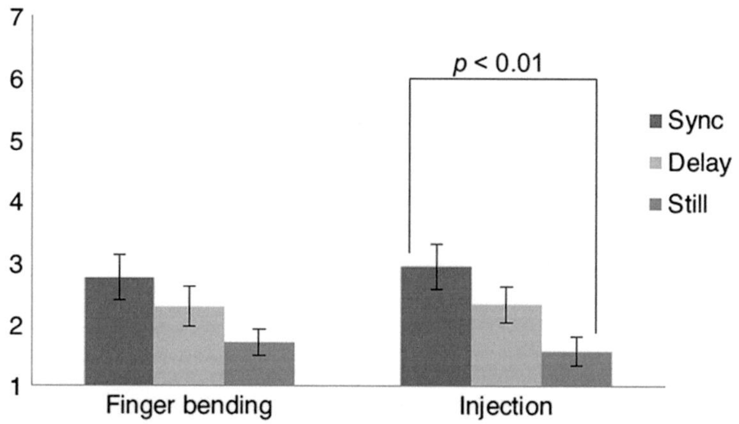

Fig. 9.25 Results of questionnaire analysis (*left*, Did it feel as if your finger was being bent?; *right*, Did it feel as if your hand was being injected?)

injection (sync condition > still condition, $p < 0.05$). Figure 9.24 shows the average with standard error and the results of multiple comparisons of Tukey HSD.

We also performed statistical analysis of the participants' self-reports. As a result of one-way ANOVA, no significant difference was confirmed in finger bending ($F(2) = 2.88$, $p = 0.06$), but a significant difference was confirmed in injection ($F(2) = 5.25$, $p < 0.01$). Multiple comparisons with Tukey HSD showed that a significant difference was confirmed only between the sync and still conditions in injection (sync condition > still condition, $p < 0.01$). Figure 9.25 shows the average with standard error and the results of multiple comparisons of Tukey HSD.

The results have verified one of the hypotheses. As for the injection stimuli, we found significant differences between the sync and still conditions, as they were revealed by both the questionnaire and SCR. In both measures, the responses under the sync condition were significantly larger than those under the still condition. Thus, hypothesis 1 (body ownership is transferred by watching scenes where the geminoid is synchronized with an operator) is confirmed. That is, we have confirmed that geminoid teleoperation induces body ownership transfer toward the geminoid body.

On the other hand, we failed to verify hypothesis 2 (in geminoid teleoperation, body ownership is transferred when the geminoid moves with delays). However, there was no significant difference between the sync and delay conditions. In the past studies on RHI, the degree of body ownership transfer was reduced by a time difference between tactile and visual stimuli (Armel and Ramachandran 2003; Botvinick 1998). Shimada et al. also showed that when the time difference became larger than 300 ms, participants had a significantly low degree of RHI effect (Shimada et al. 2009). Our result that no significant difference exists between the sync and delay (1 s) conditions may suggest that the mechanism of the geminoid body ownership transfer might be different from that of RHI body ownership transfer. In this experiment, the delay was longer than it is during usual teleoperation because we set the same delay time as in related studies. This delay might influence our result, and hence we need more verification.

A significant difference was confirmed in injection. On the other hand, no significant difference was confirmed in the questionnaire and SCR for finger bending. How can we account for this difference? First, the operation part of the body differed from the stimulus part of the geminoid; because the participants operated the geminoid's arm by moving their own arm, body ownership might be transferred to the geminoid's hand and arm but not transferred to its fingers. This might be why the participants did not react to finger bending. Second, there was a potential problem in the finger bending images we used for the experiment; because the arm position in the first picture of finger bending differed from that in the last picture, the participants saw the scenes where the arm position shifted instantaneously when the image was switched to the stimulus. This might have an unintended influence on the participants. On the other hand, we had no defective injection images.

This first point raises two interesting questions: Is body ownership transferred only to the operation part of the body? Is body ownership transferred to a large part of the body? The current teleoperation interface of the geminoid limits operator movements. If we could extend body ownership to other parts than the operation part of the body, we could improve operability.

The appearance of the operated device may influence the extent of body ownership transfer, as Pavani's study suggests (Pavani 2000). Is body ownership transfer during teleoperation only caused by androids with very humanlike appearance? Is it caused by teleoperating humanoid robots with robotic appearances or industrial robots? Future work will answer these questions.

9.5.4 Body Ownership Transfer with a Brain-Machine Interface

Since RHI was found, many researchers have studied conditions under which the illusion of body ownership transfer can be induced (Ehrsson 2007; Petkova and Ehrsson 2008).[4] In these studies, the feeling of owning a nonbody object was mainly caused by the manipulation of sensory inputs (vision, touch, and proprioception). The correlation of at least two channels of sensory information was necessary to cause the illusion. The illusion was passively evoked by synchronized visuo-tactile (Botvinick 1998) or tactile-proprioceptive (Ehrsson et al. 2005) stimulation, or it was evoked by synchronized visuo-proprioceptive stimulation in voluntarily performed action (Walsh et al. 2011).

However, the question remains whether the body ownership illusion can be induced without the correlation of multiple sensory channels. We, the author and his colleagues, are specifically interested in the role of sensory inputs in evoking motion-involved illusions. Such illusions are induced by triggering a sense of agency toward the action of a fake body. In order to trigger it, at least two afferent signals (vision and proprioception) must be integrated with efferent signals for a coherent self-presentation. Walsh et al. recently discussed the contribution of proprioceptive feedback in the inducement of the ownership illusion of an anesthetized moving finger (Walsh et al. 2011). They focused on the exclusive role of sensory receptors in muscles by eliminating the cutaneous signals from skin and joints. In contrast to this study, we hypothesize that even in the absence of proprioceptive feedback, only the match between efferent signals and visual feedback can trigger a sense of agency toward the robot's motion and therefore induce the illusion of robot body ownership.

In this study, we employ a brain-machine interface (BMI) system to translate the operator's thoughts into robot's motions and removed the proprioceptive updates of real motions from the operator's sensations. BMIs are developed primarily as a promising technology for future prosthetic limbs. Then, the incorporation of the BMI device into a patient's body representation is important for the development of BMIs. A study of monkeys has shown that bidirectional communication in a brain-machine-brain interface contributes to the incorporation of a virtual hand in a primate's brain circuitry (O'Doherty et al. 2011). Such a closed-loop multichannel BMI delivers visual feedback and multiple sensory information, such as cortically micro-stimulated tactile or proprioceptive-like signals. This kind of BMI is regarded as a future generation prosthetics that feels and acts like human bodies (Lebedev and Nicolelis 2006).

[4]The material on body ownership transfer and brain-machine interface here is adopted and modified from Alimardani et al. (2003), thanks to Science Publishing Group.

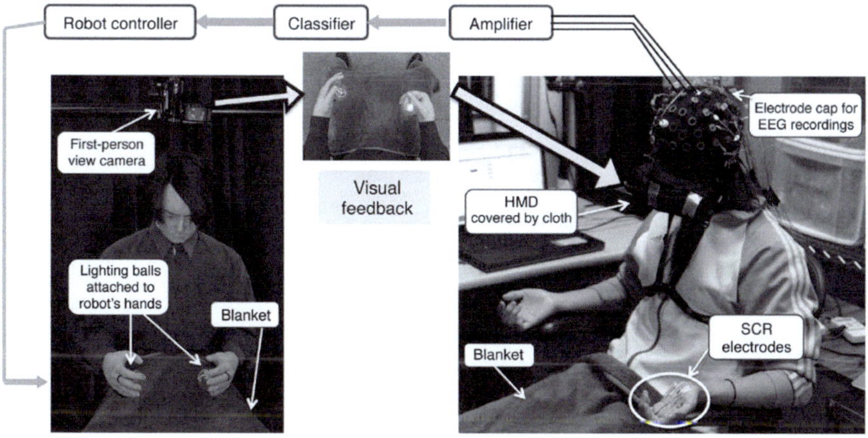

Fig. 9.26 The BMI experiment setup

Unfortunately, the design of the BMI demands invasive approaches, and they are risky and costly for human subjects. We must explore at the level of experimental studies how noninvasive operation can incorporate humanlike limbs into body presentation.

Toward this goal, we conducted experiments: subjects performed a set of motor imagery tasks of right or left grasp, and their online EEG signals were classified into two classes of right or left hand motions performed by a robot. During teleoperation, subjects wore a head-mounted display and watched real-time first-person images of the robot's hands, as shown in Fig. 9.26.

The feeling of body ownership aroused in the subjects was evaluated by subjective assessments and physiological reactions to an injection into the robot's body at the end of the teleoperation session.

9.5.5 The Methods for the BMI Experiments

We selected 40 university students as participants: 26 males and 14 females. Their average of age was 21.1 years old (standard variation was 1.92 years old), and 38 were right handed.

The cerebral activities of the subjects were recorded by biosignal amplifiers. The subjects wore an electrode cap, and 27 EEG electrodes were installed over their primary sensorimotor cortex, as shown in Fig. 9.26.

The electrode placement was based on the 10–20 system. The reference electrode was placed on the right ear and the ground electrode was on the forehead. The acquired data were processed online under Simulink/MATLAB for real-time parameter extraction. This process included band-pass filtering between 0.5 and

30 Hz, sampling at 128 Hz, cutting off artifacts by a notch filter at 60 Hz, and adopting common spatial pattern (CSP) algorithm to discriminate event-related desynchronization (ERD) and event-related synchronization (ERS) patterns associated with the motor imagery task (Neuper et al. 2006). Results were classified with weight vectors; they weighed each electrode on the basis of its importance for the discrimination task and suppressed the noise in individual channels by using the correlations between neighboring electrodes. During a right or left image movement, the decomposition of the associated EEG led to a new time series (it was optimal for the discrimination of two populations). The patterns were designed in such a way that the signal from the EEG filtering with CSP had maximum variance for the left trials and minimum variance for the right trials and vice versa. This way, the difference between the left and right populations was maximized, and the only information contained in these patterns was about where the EEG variance most fluctuated during the comparisons between the two conditions. Lastly, when the discrimination between left and right imaginations was made, the classification block outputted a linear array signal in the range of [21, 1], where 21 denotes the extreme left and 1 denotes the extreme right.

The participants imagined a grasp or squeeze motion by their own hand. In the training and experiment sessions, a visual cue specified the timing of a grasp or squeeze motion, as well as which hand they were supposed to have an image of.

The participants practiced for the motor imagery task of moving a feedback bar on a computer screen to the left or right. They sat in a comfortable chair in front of a 15-in. laptop computer and remained motionless. The first run consisted of 40 trials without feedback. They watched a cue of an arrow randomly pointing to the left or right and imagined a gripping or squeezing motion for the corresponding hand. Each trial took 7.5 s, starting with the presentation of a fixation cross on the display. An acoustic warning "beep" was given 2 s later. An arrow pointing to the left or right was shown for 3–4.25 s. Then, the participants imagined a hand motion in the direction specified by the arrow. They kept the image until the screen content was erased (7.5 s). A next trial started after a short pause. The recorded brain activities in the non-feedback run were used to set up a subject-specific classifier for the following feedback runs. In the feedback runs, the participants performed similar trials; however, after they imagined an image, the classification result was shown on the horizontal feedback bar on the screen. The task in the feedback run was to have an image immediately after the appearance of an arrow and extend the feedback bar in the specified direction as long as possible. Each of the feedback and non-feedback runs consisted of 40 randomly presented trials with 20 trials per class (left/right).

The participants performed two training sessions with feedback until they became familiar with the motor imagery task. We recorded their performances during each session to evaluate their improvement. At the end of the training sessions, most participants reached a performance of 60–100 %.

The participants wore a head-mounted display through which they had a first-person view of the robot's hands. Two balls with a light were placed in front of the robot's hands to simplify the imagery task during the experiment sessions. The participants received visual cues when the light on the right or left ball was randomly turned on and then imagined grasping by their corresponding hand. The classifier detected two classes of results (right or left) from the EEG patterns and sent a motion command to the robot's hand. Identical blankets were laid on both the robot's and subject's legs, so that the background views of the robot and subject bodies were identical.

We attached SCR electrodes to each participant's hand and measured the physiological arousal during each session. A bio-amplifier recording device with a sampling rate of 1000 Hz was used for the SCR measurements.

The participants rested their hands with palms up on the chair arms, and SCR electrodes were placed on the thenar and hypothenar eminences of their left palm. Since the robot's hands were designed to be bent inward for grasping motion, the participants tended to turn their hand inward so that it resembled the robot's hand. This might give the participants the space and comfort to perform unconscious gripping motions during the task and complicate the arrangement of SCR electrodes. To avoid this problem, we asked the participants to keep their hands and elbows motionless on the chair arms with their palms up.

The participants practiced operating the robot's hand by motor imagery in one session and then performed two test sessions. All sessions consisted of 20 trials. The test sessions were randomly assigned either a *still* condition or a *match* condition. In the still condition, the robot's hands did not move at all, even though the participants performed imagery tasks with a cue stimulus and expected an expectation of robot's motion. In the match condition, the robot's hands performed a grasping motion but only in those trials whose classification results were correct and identical as the cue. If the subjects made a mistake during a trial, the robot's hands did not move.

At the end of each test session, a syringe was injected into the thenar muscle of the robot's left hand, which was the same hand on which the SCR electrodes had been placed, as shown in Fig. 9.27. We slowly moved the syringe toward the robot's hand: it took about 2 s after the injector appeared in the participant's view that it touched the robot's skin. Immediately after the injection, the session was terminated, and the participants were orally asked the following two questions:

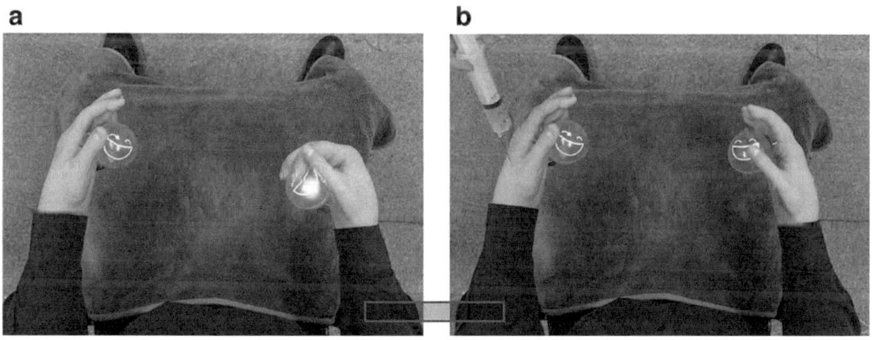

Fig. 9.27 Participant's view in HMD. (**a**) Robot's right hand grasped the lighted ball. (**b**) Robot's left hand received injection

(Q1) When the robot's hand was given a shot, did it feel as if your own hand was being injected?
(Q2) Throughout the session while you were performing the task, did it feel as if the robot's hands were your own hands?

They scored each question on a seven-point Likert scale, where 1 was "didn't feel at all" and 7 was "felt very strongly."

SCR measurements: the peak value of the response to the injection was selected as a reaction value. Generally, SCRs start to rise 1 or 2 s after a stimulus and end 5 s after it. The moment at which the syringe appeared in the participant's view was selected as the starting point of evaluation, because some participants reacted to the syringe itself even before it was inserted into the robot's hand as a result of the body ownership illusion. Therefore, SCR peak values were sought within an interval of 6 s: 1 s after the syringe appeared in the participant's view (1 s before it was inserted) to 5 s after the injection was actually made.

We averaged and compared the acquired scores and the SCR peak values for each condition within subjects. Statistical analysis was carried out by paired *t*-test. A significant difference between the two conditions was revealed in Q1 and Q2 (Match > Still, $p < 0.001$), and in the SCR responses (Match > Still, $p < 0.01$).

9.5.6 The Results of the BMI Experiments

Each session was conducted either in the still condition or in the match condition.

Still condition: The robot's hands did not move at all, even though subjects performed the imagery task and tried to operate the hands (this was the control condition).
Match condition: On the basis of the classification results, the robot's hands performed a grasping motion but only when the result was correct.

If the participant missed a trial, the hands did not move. The participants were unaware of the condition setups. They were told that the good performance of the task would produce a robot motion. To help imagination and give a visual cue for the motor imagery tasks, two balls were placed in front of the robot's hands, and the balls randomly lighted to indicate which hand the subjects were required to move (Fig. 9.27a). At the end of each test session, a syringe was inserted into the thenar muscle of the robot's left hand (Fig. 9.27b). Immediately after the injection, the session was terminated, and the participants were asked questions Q1 and Q2.

The acquired scores for each condition were averaged and compared within subjects by paired *t*-test (Fig. 9.28a). The Q1 results were significant between the match (M = 4.38, SD = 1.51) and still (M = 2.83, SD = 1.43) conditions [Match > Still, $p = 0.0001$, $t(39) = -4.33$]. Similarly, there was a significant difference in the Q2 scores between the match (M = 5.15, SD = 1.10) and still (M = 2.93, SD = 1.25) conditions [Match > Still, $p = 2.75 \times 10^{-12}$, $t(39) = -9.97$].

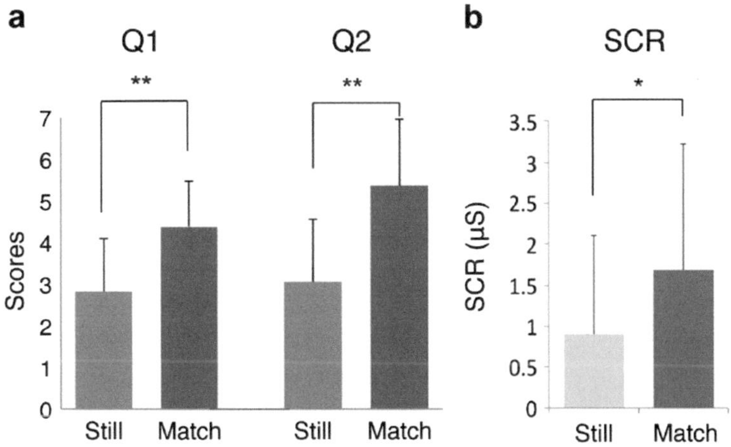

Fig. 9.28 The experimental results of the questionnaire (**a**) and SCR (**b**)

We physiologically measured the body ownership illusion not only by the self-assessment but also by recording the skin conductance responses (SCR). The peak response value within a 6-s interval was selected as the SCR reaction value. The SCR results were significant between the match ($M = 1.68$, $SD = 1.98$) and still ($M = 0.90$, $SD = 1.42$) conditions [Match > Still, $p = 0.002$, $t(33) = -3.29$], although the subject responses were spread over a large range of values (Fig. 9.28b).

9.6 Discussion

From both the questionnaire and SCR results, we can conclude that the operator's reactions to a painful stimulus (injection) were significantly stronger in the match condition in which the robot's hands followed the operator intentions. This reaction is evidence for the body ownership illusion and verifies our hypothesis. Body ownership transfer to the robot's moving hands can be induced without proprioceptive feedback from the operator's actual movement. This is the first report on the body ownership transfer to a nonbody object that is induced without integration among multiple afferent inputs. In this case, a correlation between efferent information (operator's plan of motion) and a single channel of sensory input (visual feedback of the intended motion) was enough to trigger the illusion of body ownership. Since this illusion occurs in the context of action, we estimate that the feeling or sense of ownership of the robot's hand is a modulated sense of agency generated for the robot hand motions. Although all the participants were perfectly aware that the hands they were seeing through the HMD were nonbody humanlike objects, the explicit sense that "I am the one causing the motions of these hands" and the lifelong experience of performing motions for their own bodies modulated the sense of body ownership toward the robot's hands and provoked the illusion.

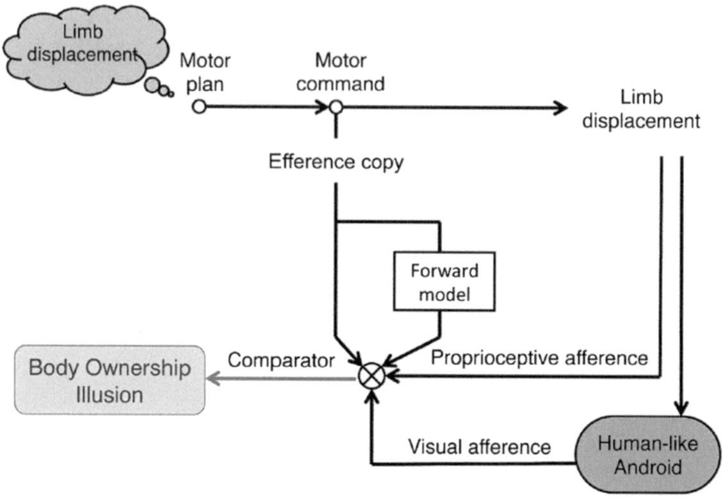

Fig. 9.29 Body recognition mechanism

The ownership illusion for geminoids was a visuo-proprioceptive illusion due to the motion synchronization between the operator and the robot's body. The mechanism behind this illusion can be explained on the basis of a cognitive model that integrates one's body into oneself in a self-generated action (Tsakiris et al. 2007). When an operator moves his/her body and watches the robot copying the movement, the match between the motion commands and the sensory feedbacks from the motion modulates a sense of agency over the robot's actions. Here, the sensory feedbacks mean visual afference from the robot's body and proprioceptive afference from the operator's body. Then, the integration of robot's body into operator's sense of self-body occurs, as shown in Fig. 9.29. The experiments particularly target the role of proprioception in this model and show that our presented mechanism remains valid even when the proprioceptive feedback channel is blocked from being updated.

An important element associated with the occurrence of the illusion is probably the attention to the operation task. Motor imagery is a difficult process and requires increased attention and concentration. Subjects focus on picturing a grasping motion. Their selective attention to the visual feedback can play a noticeable role in filtering out such distracting information, as different sizes and textures of visible hands, the delay between the onset of motor imagery and the robot's motions, and the subconscious sense of the real hand positions.

On the other hand, the exclusively significant role of visual feedback in the process of body attribution is disputable, since the cutaneous signals of the subject's real body were not blocked. This may reflect the effect of the movement-related gating of sensory signals before and during the execution of self-generated movements. In a voluntary movement, feed-forward efference copy signals predict the expected outcome of the movement. This expectation modulates the incoming sensory input and reduces the transmission of tactile inputs (Chapman et al. 1988). This attenuation of sensory input has also been reported at the motor planning stage (Voss et al. 2008) and has been found in the neural patterns of primates prior to the actual onset of voluntary movement (Lebedev et al. 1994).

These findings support the premovement elevation of a tactile perception threshold during motor imagery tasks. Therefore, the suppression or gating of peripheral signals may have enhanced the relative salience of other inputs, which are visual signals in this case.

Although we only discuss motor-visual interaction on the basis of a previously introduced cognitive mechanism of body recognition, there is evidence that the early stages of movement preparation not only occur in the brain's motor centers but may also occur simultaneously in the spinal levels (Prut and Fetz 1999). This suggests that even in the absence of the subject's movement, further sensory circuitry may in fact be engaged in the presented mechanism at the level of motor planning.

Correspondingly, although the subjects were strictly prohibited from performing grasp motions by their own hands, it is probable that some of them did occasionally contract their upper arm muscles. This possibility prompts more research into the complete cancelation of proprioceptive feedback. In the future, further experiments with EMG recordings are required to improve the consistency of our results by excluding participants whose performance involves muscle activity.

Lastly, from the observations of this experiment and many other studies, we conclude that with regard to illusions of body transfer, the congruence between two channels of information, whether efferent or afferent, is sufficient to integrate a nonbody part into one's own body, no matter in what situation the body transfer experience occurs. The integration of two sensory modalities from nonbody and body parts is necessary for passive or externally generated body transfer experiences. However, since efferent signals play a critical role in the recognition of one's own voluntary motions, their congruence with only a single channel of visual feedback from nonbody part motions was adequate to override the internal mechanism of body ownership.

9.7 Concluding Remarks

This chapter has discussed android science as an interdisciplinary framework bridging robotics and cognitive science. Android science is expected to be a fundamental research area in which the principles of human-human communications and human-robot communications are studied.

In developing the geminoid, one of the important but unmentioned topics is to study sonzai-kan or human existence/presence; the study on this topic will extend the framework of android science. The scientific part of android science must answer questions concerning how humans recognize human existence/presence. On the other hand, the technological part of android science must develop a teleoperated android that adequately works and has human or humanlike existence/presence.

Exercises

1. Discuss how we should balance teleoperated functions and automated functions in order to realize the geminoid that can work as the avatar of the operator.
2. It is of merit if the geminoid can generate movements autonomously according to operator's voice, since it is difficult to control them remotely. Consider the voice-based methods for generating movements, such as head movements, eye blinking, and body movements.
3. Design a new psychological experiment on humanlikeness by using androids.
4. Discuss practical applications of androids and geminoids in the next decade.

References

Alimardani, M., Nishio, S., Ishiguro, H.: Humanlike robot hands controlled by brain activity arouse illusion of ownership in operators. Sci. Rep. **3**, 2396 (2013)

Armel, K., Ramachandran, V.: Projecting sensations to external objects: evidence from skin conductance response. Proc. Biol. Sci. **270**(1523), 1499–1506 (2003)

Botvinick, M.: Rubber hands 'feel' touch that eyes see. Nature **391**(6669), 756 (1998)

Brooks, R.: Intelligence without representation. Artif. Intell. **47**, 139–159 (1991)

Chapman, C., Jiang, W., Lamarre, Y.: Modulation of lemniscal input during conditioned arm movements in the monkey. Exp. Brain Res. **72**, 316–334 (1988)

Ehrsson, H.: The experimental induction of out-of-body experiences. Science **317**, 1048 (2007)

Ehrsson, H., Holmes, N., Passingham, R.: Touching a rubber hand: feeling of body ownership is associated with activity in multisensory brain areas. J. Neurosci. **25**, 10564–10573 (2005)

Harnad, S.: The symbol grounding problem. Phys. D **42**, 335–346 (1990)

Hashimoto, T., Hiramatsu, S., Kobayashi, H.: Development of face robot for emotional communication between human and robot. In: Proceedings of the IEEE International Conference on Mechatronics and Automation, (2006)

Hollan, J., Hutchins, E., Kirsh, D.: Distributed cognition: toward a new foundation for human-computer interaction research. ACM Trans. Comput. Hum. Interact. **7**(2), 174–196 (2000)

Home page of the Loebner Prize in artificial intelligence, "The first Turing Test," http://www.loebner.net/Prizef/loebner-prize.Html

Ikeda, T., Ishida, T., Ishiguro, H.: Framework of distributed audition. In: Proceedings of the 13th IEEE International Workshop of Robot and Human Interactive Communication (ROMAN), pp. 77–82, (2004)

Ishiguro, H.: Scientific issues concerning androids. Int. J. Robot. Res. **26**(1), 105–117 (2007)

Ishiguro, H., Nishio, S.: Building artificial humans to understand humans. J. Artif. Organs **10**(3), 133–142 (2007)

Ishiguro, H., Nishimura, T.: VAMBAM: view and motion based aspect models for distributed omnidirectional vision systems. In: Proceedings of the International of the Joint Conference on Artificial Intelligence (IJCAI), pp. 1375–1380, (2001)

Ishiguro, H.: Distributed vision system: a perceptual information infrastructure for robot navigation. In: Proceedings of the International Joint Conference on Artificial Intelligence (IJCAI), pp. 36–41, (1997)

Ishiguro, H., Ono, T., Imai, M., Maeda, T., Kanda, T., Nakatsu, R.: Robovie: an interactive humanoid robot. Int. J. Ind. Robot. **28**(6), 498–503 (2001)

Ishiguro, H.: Toward interactive humanoid robots: a constructive approach to developing intelligent robot. In: Proceedings of the 1st International Joint Conference on the Autonomous Agents and Multiagent Systems, Invited talk, Part 2, pp. 621–622, (2002)
Itakura, S.: Gaze following and joint visual attention in nonhuman animals. Jpn. Psychol. Res. **46**, 216–226 (2004)
Kanda, T., Ishiguro, H., Ishida, T.: Psychological analysis on human-robot interaction. In: Proceedings of IEEE International Conference on Robotics and Automation (ICRA), pp. 4166–4171, (2001)
Kanda, T., Ishiguro, H., Imai, M., Ono, T.: Development and evaluation of interactive humanoid robots. Proc. IEEE **92**(11), 1839–1850 (2004)
Kilner, J., Paulignan, Y., Blakemore, S.: An interference effect of observed biological movement on action. Curr. Biol. **13**, 522–525 (2003)
Lang, P.J., Greenwald, M.K., Bradley, M.M., Hamm, A.O.: Looking at pictures: affective, facial, visceral, and behavioral reactions. Psychophysiology **30**, 261–273 (1993)
Lebedev, M., Nicolelis, M.: Brain-machine interfaces: past, present and future. Trends Neurosci. **29**, 536–546 (2006)
Lebedev, M., Denton, J., Nelson, R.: Vibration-entrained and premovement activity in monkey primary somatosensory cortex. J. Neurophysiol. **72**, 1654–1673 (1994)
MacDorman, K., Minato, T., Shimada, M., Itakura, S., Cowley, S.J., Ishiguro, H.: Assessing human likeness by eye contact in an android testbed. In: Proceedings of Annual Meeting of the Cognitive Science Society, (2005)
McCarthy, A., Lee, K., Muir, D.: Eye gaze displays that index knowing, thinking and guessing. In: Proceedings of the Annual Conference on American Psychological Society, (2001)
Minato, T., Shimada, M., Itakura, S., Lee, K., Ishiguro, H.: Evaluating the human likeness of an android by comparing gaze behaviors elicited by the android and a person. Adv. Robot. **20**, 1147–1163 (2006)
Mori, M.: Bukimi no tani (the uncanny valley). Energy **7**, 33–35 (1970)
Neuper, C., Muller-Putz, G., Scherer, R., Pfurtscheller, G.: Motor imagery and EEG-based control of spelling devices and neuroprostheses. Prog. Brain Res. **159**, 393–409 (2006)
Nishio, S., Ishiguro, H., Hagita, N.: Geminoid: teleoperated android of an existing person. In: de Pina Filho, A. (ed.) Humanoid Robots: New Developments. I-Tech Education and Publishing, Vienna (2007)
Nishio, S., Watanabe, T., Ogawa, K., Ishiguro, H.: Body ownership transfer to teleoperated android robot. In: Paper presented at the International Conference on Social Robotics, pp. 398–407, (2012)
O'Doherty, J., et al.: Active tactile exploration using a brain-machine-brain interface. Nature **479**, 228–231 (2011)
Pavani, F.: Visual capture of touch: out-of-the-body experiences with rubber gloves. Psychol. Sci. **11**(5), 353–359 (2000)
Perani, D., Fazio, F., Borghese, N., Tettamanti, M., Ferrari, S., Decety, J., Gilardi, M.: Different brain correlates for watching real and virtual hand actions. NeuroImage **14**, 749–758 (2001)
Perlin, K.: Real time responsive animation with personality. IEEE Trans. Vis. Comput. Graph. **1**(1), 5–15 (1995)
Personal robot PaPeRo, NEC Co. (Online). Available http://www.incx.nec.co.jp/robot/PaPeRo/english/p index.html
Petkova, V., Ehrsson, H.: If I were you: perceptual illusion of body swapping. PLoS One **3**, e3832 (2008)
Prut, Y., Fetz, E.: Primate spinal interneurons show pre-movement instructed delay activity. Nature **401**, 590–594 (1999)
Shimada, S., Fukuda, K., Hiraki, K.: Rubber hand illusion under delayed visual feedback. PLoS One **4**(7), e6185 (2009)
Tsakiris, M.: My body in the brain: a neurocognitive model of body-ownership. Neuropsychologia **48**(3), 703–712 (2010)

Tsakiris, M., Haggard, P., Frank, N., Mainy, N., Sirigu, A.: A specific role for efferent information in self-recognition. Cognition **96**, 215–231 (2007)

Turing, A.: Computing machinery and intelligence. Mind **59**, 433–460 (1950)

Voss, M., Ingram, J., Wolpert, D., Haggard, P.: Mere expectation to move causes attenuation of sensory signals. PLoS One **3**, e2866 (2008)

Walsh, L., Moseley, G., Taylor, J., Gandevia, S.: Proprioceptive signals contribute to the sense of body ownership. J. Physiol. **589**, 3009–3021 (2011)

Index

M. Kasaki et al. (eds.), *Cognitive Neuroscience Robotics A*,
DOI 10.1007/978-4-431-54595-8

MIX
Papier aus verantwortungsvollen Quellen
Paper from responsible sources
FSC® C105338

If you have any concerns about our products,
you can contact us on
ProductSafety@springernature.com

In case Publisher is established outside the EU,
the EU authorized representative is:
Springer Nature Customer Service Center GmbH
Europaplatz 3, 69115 Heidelberg, Germany

Printed by Libri Plureos GmbH
in Hamburg, Germany